Current Topics in Microbiology and Immunology

Volume 440

The reviews series *Current Topics in Microbiology and Immunology* publishes cutting-edge syntheses of the latest advances in molecular immunology, medical microbiology, virology and biotechnology. Each volume of the series highlights a selected timely topic, is curated by a dedicated expert in the respective field, and contains a wealth of information on the featured subject by combining fundamental knowledge with latest research results in a unique manner.

2020 Impact Factor: 4.291, 5-Year Impact Factor: 5.110

2020 Eigenfactor Score: 0.00667, Article Influence Score: 1.480

2020 Cite Score: 7.7, h5-Index: 38

Robert Garry
Editor

Lassa Fever: Epidemiology, Immunology, Diagnostics, and Therapeutics

Responsible Series Editor: Michael Oldstone

Springer

Editor
Robert Garry
Department of Microbiology
and Immunology
Tulane University School of Medicine
New Orleans, LA, USA

Zalgen Labs, LLC
Frederick, MD, USA

ISSN 0070-217X ISSN 2196-9965 (electronic)
Current Topics in Microbiology and Immunology
ISBN 978-3-031-35806-7 ISBN 978-3-031-35807-4 (eBook)
https://doi.org/10.1007/978-3-031-35807-4

This Springer imprint is published by the registered company Springer Nature Switzerland AG
The registered company address is: Gewerbestrasse 11, 6330 Cham, Switzerland

Preface

During my first semester of graduate school at the University of Texas at Austin in Fall 1974, I was faced with a choice about my graduate career. I was certain that I wanted to be a virologist after undergraduate work on viruses in the lab of Dr. Wayne Magee at Indiana State University. However, my advisors at UT strongly suggested that I pursue work on bacterial viruses. With my best interests at heart, I was steered away from the relatively underdeveloped field of animal virology. Reverse transcriptase had just been discovered by Temin and Baltimore in 1970. While there was much fanfare with this discovery, no one had yet heard of human retroviruses, the human T-lymphotropic viruses (HTLV-I and HTLV-II) and human immunodeficiency virus 1 and 2 (HIV-1 and HIV-2). Neither hepatitis C nor Ebola virus had been discovered. It would not be for two more years that Dr. Harold zur Hausen would publish his hypothesis, which was later confirmed, that human papillomaviruses cause most cervical cancers. Many of the major biological discoveries, DNA as the hereditary material, the molecular gymnastics of DNA synthesis, the role mRNA in protein synthesis, and the powerful impact of mutations had come from the study of bacterial viruses. I was nearly convinced to pick a lab of one of several of the skilled, well-established, and well-funded microbial geneticists at UT. Because of a plethora of excellent possibilities, I delayed my choice.

As fate would have it, I picked up a book **Fever!: The Hunt for a New Killer Virus** on the new release shelf in the Science section of the UT library. I must have been attracted to the cover illustration, which was not unlike some of my beloved pulp fiction/comic book covers (Fig. 1). Little did I realize that the author, John Grant Fuller, had honed his writing and investigatory skills by focusing mainly on extraterrestrials and the supernatural. It was of no matter. The story that Fuller told here was true, but no less dramatic than his other works that required a suspension of disbelief. The narrative began in January 1969 when Laura Wine a missionary nurse working in Lassa, a remote village in northeastern Nigeria contracted a severe illness. Nurse Wine was air-lifted to Bingham Hospital in the city of Jos approximately 300 km from Lassa, but died 30 h after admission. Within a week Charlotte Shaw one of the Bingham nurses had contracted the disease dying on the 11th day of her illness. A third nurse, Lily (Penny) Pinneo, who had attended both women and assisted in the autopsy of Shaw, also contracted the illness. Nurse Pinneo was evacuated via Lagos to New York Presbyterian Hospital. After a lengthy illness and recovery she survived.

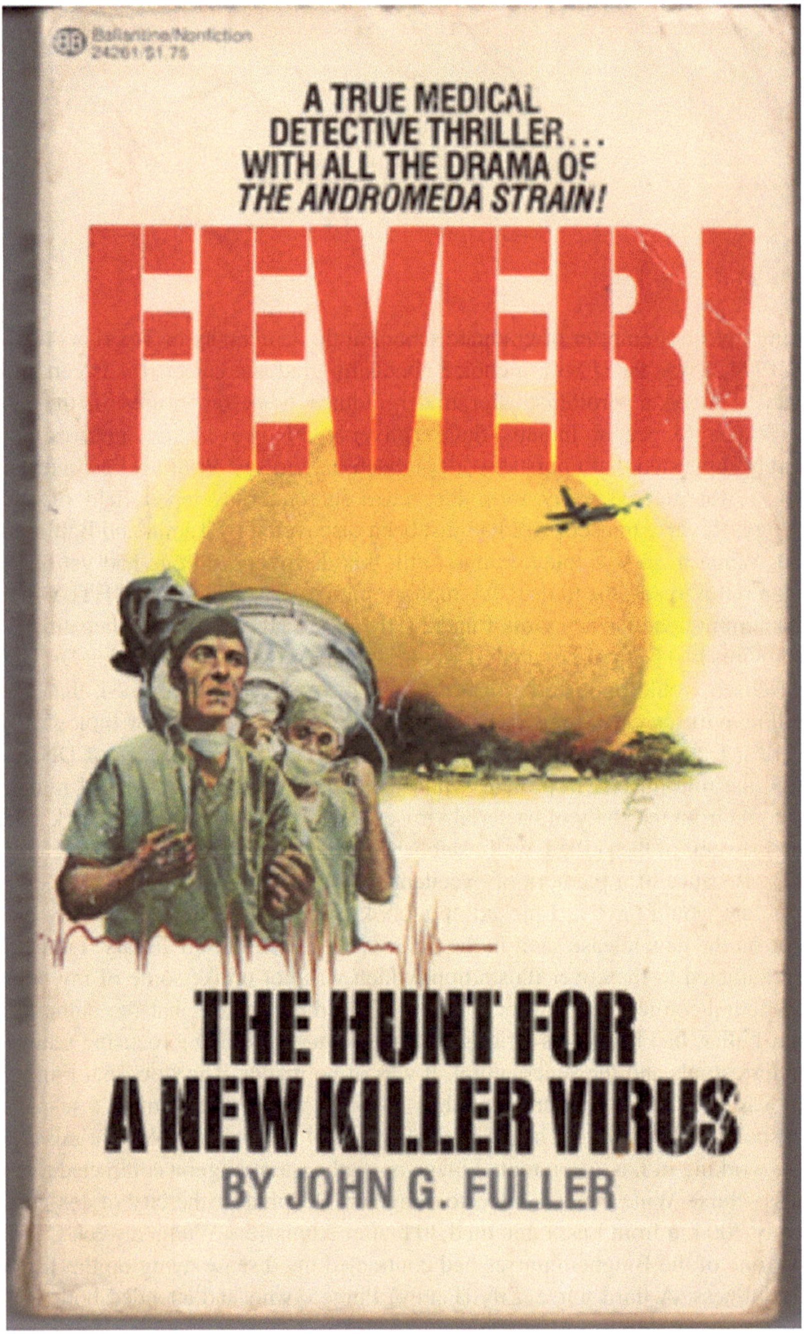

Fig. 1 Cover from Fever! The hunt for a new killer virus by John G. Fuller. Reproduced with permission from Penguin Random House LLC.

Serum from the three missionary nurses drawn in Africa during the acute phase of their illnesses was transported on ice to New York. Blood was also drawn by John Frame on March 6 as Nurse Pinneo arrived in New York and was forwarded with the other specimens to the Yale Arbovirus Research Unit (YARU) in New Haven, Connecticut. The March 6 specimen from Pinneo was the first to show the presence of cytopathic effects (CPE) in cell culture. Two YARU researchers were infected with LASV during these initial studies. Juan Roman, a research technician, died. The other researcher, Dr. Jordi Casals an eminent virologist, received blood donated from Pinneo as passive immunotherapy and survived. A novel arenavirus was isolated from the blood of Shaw, Pinneo, and Casals. The isolate from Pinneo was designed as the type strain, now known as Lassa virus (LASV). The Pinneo strain from northern Nigeria remains one of the few LASV lineage I isolates.

Reading this account of virus discovery hooked me on animal viruses. I aspired to do work on animal and perhaps human viruses with the rigor of the phage groups. I chose to work on Sindbis virus, an alphavirus, with the newest member of the UT Microbiology Department Dr. Marilynn Waite. With a modest budget, we conducted basic research on CPE mediated by Sindbis virus. I stayed on at UT to do an individual National Institutes of Health post-doctoral fellowship with Dr. Henry "Hank" Bose on reticuloendotheliosis virus (REV), an immunosuppressive retrovirus of birds. In 1983, I had the opportunity to establish my own lab at Tulane University in New Orleans. The chair of the department Dr. Author Gottlieb, an immunologist, was interested in a new disease characterized by profound immunosuppression. He provided me with serum samples from a New Orleans patent with the disease, who had been sexually exposed in New York City to cases of the newly named acquired immunodeficiency syndrome (AIDS). I exposed peripheral blood mononuclear cells to the patient's serum and after a day or two saw striking CPE, including syncytial cell formation. The story from **Fever!** of Jordi Casals infected in his lab with an unknown virus crossed my mind working on these early cultures. This is the one reason I wore gloves and worked carefully in a class II hood we had moved into "isolation" in what was formerly a broom closet. Luckily, I did not infect myself with the virus, which would soon be described by the groups of Montagnier and Gallo and later named HIV.

My work with Hank Bose on REV was a fortuitous head start to working on HIV, the newly described immunosuppressive human retrovirus. I began investigating the mechanisms of HIV-induced CPE in T-cells. Visiting the lab of Dr. Suraiya Rasheed at the University of Southern California I discovered how to grow HIV in persistently-infected cultures to large quantities. I was part of a collaborative team headed by Dr. Steve Alexander at Biotech Laboratories that developed first generation HIV immunoassays (western blots) that we used to retrospectively identify an early AIDS case from 1968, the year before the first description of Lassa fever. I would continue my work on HIV for several decades. When the 9–11 terrorist attack and the Amerithrax incident highlighted the need for advanced diagnostics for biodefense and emerging pathogens, I assembled a team of industry and academic scientists committed to developing immunodiagnostics for Lassa fever. Later, other scientists joined us to work on LASV and other emerging pathogens in remarkable collaborative groups, including the Viral Hemorrhagic Fever Consortium (VHFC), the Africa Center of Excellence for the Genomics of Infectious Diseases (AECEGID), the Viral Immunotherapeutics Consortium (VIC), and the Center for Viral Systems Biology (CViSB). In this book, we summarize our work and that of many other talented investigators. We also look ahead to the future of Lassa research and perhaps to the beginning of the end of Lassa fever as a major public health threat to the people of West Africa.

New Orleans, USA
August 2019

Robert Garry, Ph.D.
rfgarry@tulane.edu

Contents

50 Years of Lassa Fever Research

Robert F. Garry

Contents

Abstract Lassa fever was first described as a clinical entity fifty years ago. The causative agent Lassa virus was isolated from these first known cases. This chapter reviews the key publications on Lassa fever research that appeared in the scientific literature at that time and over the ensuing decades.

R. F. Garry
Department of Microbiology and Immunology, Tulane University School of Medicine, 1430 Tulane Avenue, New Orleans, LA 70118, USA

R. F. Garry (✉)
Zalgen Labs, LLC, 20271 Goldenrod Lane, Suite 2083, Germantown, MD 20876, USA
e-mail: rfgarry@tulane.edu

Current Topics in Microbiology and Immunology (2023) 440: 1–22
https://doi.org/10.1007/82_2020_214

Published Online: 27 May 2020

1 Introduction

50 years ago as of this writing Lassa virus (LASV) was discovered and shown to be the causative agent of Lassa fever (Buckley et al. 1970; Frame et al. 1970; Pinneo and Pinneo 1971) Lassa fever is endemic in the countries of Nigeria, Sierra Leone, Guinea, and Liberia. Sporadic cases have been reported in other West African countries, with recent outbreaks in Togo and Benin (Shaffer et al. 2014; Andersen et al. 2015; WHO 2017; Manning et al. 2015; ECDC 2016) LASV isolates display high genetic diversity (Andersen et al. 2015; Bowen et al. 2000), with an expanding number of lineages and sub-lineages (Manning et al. 2015; Ehichioya et al. 2019; Whitmer et al. 2018). The main reservoir of LASV is *Mastomys natalensis*, a peridomestic rodent that is abundant across a wide range in sub-Saharan Africa (Monath et al. 1974b). LASV is transmitted vertically to the offspring of infected rodents and horizontally to humans via urine or feces or during the preparation of the rodent for food. Additional rodent reservoirs or intermediate hosts have been described and may vary by lineage (Olayemi et al. 2016b; Yadouleton 2019). Currently, the only available treatment is off-label use of the nucleoside drug ribavirin (Eberhardt et al. 2019). Progress has been made toward the development of an immunotherapeutic (Mire et al. 2017; Cross et al. 2016b) and a small molecule drug (Madu et al. 2018). While there is no approved Lassa fever vaccine, there are new initiatives to address this urgent public health need (Burki 2017). This chapter will review the presentation of Lassa fever research in the scientific literature.

2 Overview of the Lassa Literature

A PubMed search for “Lassa” on August 18, 2019, returned 1438 entries mentioning the virus or disease. By comparison “Ebola” returned 8458 entries, while “HIV” returned 350,652. A search for “cancer” returned almost 4 million entries (3,914,091). During the 1970s, the decade in which Lassa fever was first formally described there were 178 publications on Lassa fever or Lassa virus (Fig. 1). There were 220 Lassa publications in the 1980s, most from dedicated investigators with the US Centers for Disease Prevention and Control (CDC), the Army Medical Research Institute for Infectious Diseases (USAMRIID), the Special Pathogens Reference Laboratory at Porton Down in the UK, the Byelorussian Research Institute of Epidemiology and Microbiology in the Union of Soviet Socialist Republics (USSR) or the University of Ibadan, Nigeria. The civil conflict in West Africa as well as a general reduction in support for biomedical research leads to a drop off to only 119 articles published during the 1990s. However, spurred by the surge in funding for work on biothreat agents as a result of the anthrax letters sent after the 9/11 attack, the numbers of Lassa publications grew to 258 in the first decade of the new millennium. The inadequate response to the Ebola outbreak of 2013–16 prompted an acceleration of research on all emerging viruses (Ryschon

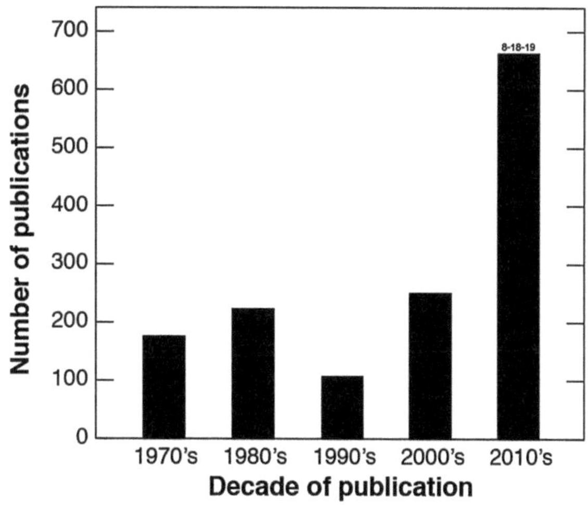

Fig. 1 Number of Lassa fever publications in the scientific literature by decade

2014). 663 articles on Lassa have been published through mid-August 2019 with more to come before the end of the decade. There have been advances in understanding the virology, ecology, molecular biology, immunology, and genomics of LASV and progress toward countermeasures including diagnostics, therapeutics, and vaccines, which we summarize below.

The widespread distribution of LASV across West Africa coupled with evidence of natural selection for LASV resistance in humans suggests that Lassa fever might be an ancient human disease (Andersen et al. 2015; Sabeti et al. 2006). Molecular dating determined that current LASV variants originated in modern-day Nigeria approximately a thousand years ago and spread to countries to the west within the last several hundred years (Andersen et al. 2015). We estimated that the time to the most common recent ancestor (tMRCA) of sampled LASV strains to be 1057 years ago for the L segment and 650 years ago for the S segment (Andersen et al. 2015). In contrast, LASV isolates in Sierra Leone have an estimated tMRCA of 150 years. A recent phylogeographic analysis of Liberian LASV genomes estimates a tMRCA of 300–350 years (Wiley et al. 2019). The scientific literature, however, contains little evidence for human infections prior to the discovery of LASV in 1969. Two missionaries were shown retrospectively to have a serological response to LASV after experiencing a febrile illness in 1952 while working in Nigeria (Fig. 2) (Bond et al. 2013). A likely outbreak of Lassa fever occurred in 1955 at Nixon Memorial Hospital in Segbwema, Sierra Leone (Fig. 2) (Rose 1956). It seems unlikely that LASV, which does not spread effectively in humans, abruptly acquired the ability to cause human disease. These retrospective examples were probably not the only human cases of Lassa fever in West Africa prior to the formal description of the disease. It is difficult to distinguish common febrile illnesses, such as malaria, typhoid fever, leptospirosis, and arbovirus diseases, from Lassa fever on the basis of clinical presentation alone. Sporadic outbreaks with high morbidity and mortality

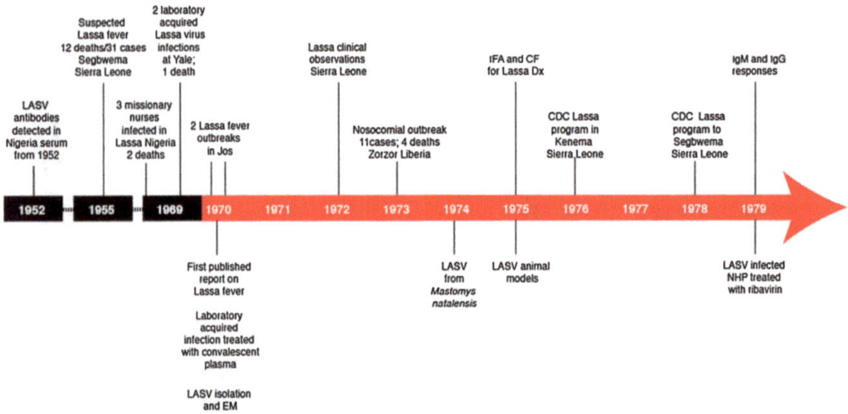

Fig. 2 Key events and scientific publications regarding Lassa fever prior to and during the 1970s

are likely to have occurred. Anecdotally, there are accounts of deaths due to what was likely Lassa fever in people living in Kenema District and surrounding areas going back to at least the 1940s (Dr. Donald S. Grant, personal communication).

3 Lassa Fever Research by Decade

3.1 The 1970s

Lassa fever was first described in 1969 (Fig. 2) (Frame et al. 1970; Buckley and Casals 1970; Speir et al. 1970). The first known patient was a Serving In Mission (SIM) nurse, who was infected while working in a hospital near Lassa, Nigeria (Pinneo and Pinneo 1971). She was transported to Bingham Hospital in the city of Jos some 300 miles away, where she died. Two other Bingham missionary nurses were infected while caring for their colleague and died. A third Bingham nurse, Lily (Penny) Pinneo was also infected, but survived after being transported to New York City (Watts 2013). Blood samples from the infected nurses were shipped to the Yale Arbovirus Research Unit (YARU) where the first Lassa laboratory research was initiated. In June 1969, two YARU researchers were infected with LASV, one fatally (Leifer et al. 1970). Dr. Jordi Casals, an eminent virologist, received blood donated from Pinneo as passive immunotherapy (Leifer et al. 1970). LASV, a novel arenavirus, was isolated from the blood of the missionary nurses and Casals (Buckley et al. 1970; Buckley and Casals 1970). The isolate from Pinneo was designed as the type strain.

In early 1970, two hospitals in Jos, Nigeria, experienced nosocomial outbreaks of Lassa fever, including Bingham Hospital (Troup et al. 1970; Carey 1972). There were 28 hospital-acquired cases with 13 deaths (46% case fatality rate, CFR)

(Troup et al. 1970; White 1972). The outbreak was investigated by a team from the Rockefeller Foundation Virus Research Laboratory in Ibadan, Nigeria, led by Dr. Don Carey who described the epidemiology, clinical features, and pathology of the disease (Troup et al. 1970; Carey 1972; Henderson et al. 1972). Dr. Jeanette Troup was tragically infected while performing an autopsy of one of the 1970 Jos cases and died from Lassa fever.

In March 1972, Dr. Thomas P. Monath a CDC research fellow at the Rockefeller laboratory in Ibadan, Nigeria, was called to investigate a Lassa fever outbreak in Zorzor, Liberia (Monath et al. 1973). Monath, who participated in an important segment of the early work in Sierra Leone, has recently written an historical review of the first 10–15 years after discovery, which is highly recommended (Monath 2019). Monath recounts how he was joined in Liberia by Nurse Pinneo, who was immune to LASV infection (Bond et al. 2013) and performed many of the high-risk procedures. As in Jos, most of the 11 cases (4 deaths, 36%) were on a single ward, in this instance the obstetrics ward. Shortly after the outbreak in Zorzor, CDC was alerted to an outbreak in Panguma, a village in the Eastern Province of Sierra Leone (Monath et al. 1974a; Fraser et al. 1974). The Panguma outbreak was different than previous outbreaks that were driven by nosocomial infections. The 63 likely cases in and around Panguma were community-acquired and had occurred over two years. It was during the investigation of these cases that Monath identified the rodent host of LASV, the multimammate mouse *Mastomys natalensis* (Monath et al. 1974b).

In 1976, the Sierra Leone Ministry of Health and Sanitation (MoHS) and CDC established Lassa fever treatment wards and research facilities in the Eastern Province (Khan et al. 2008). Three centers at Panguma, Segbwema, and Kenema became major sites for Lassa fever research throughout the 1970s and 1980s (McCormick and Fisher-Hoch 1996). The CDC Lassa Fever Project was initially headquartered in Kenema, under the leadership of Dr. Joseph McCormick and moved to Segbwema in 1978 when Dr. Patricia Webb became Director (Chu 2018). The 1970s were marked by the development of the first Lassa fever diagnostics (Frame et al. 1979) including complement fixation and immunofluorescence assays. (Fabiyi and Tomori 1975; Arnold and Gary 1977; Wulff and Johnson 1979) Later in the decade, the first animal models were described (Kiley et al. 1979; Stephen and Jahrling 1979) and there was the first demonstration in nonhuman primates (NHPs) that ribavirin might be a treatment option for Lassa fever (Stephen and Jahrling 1979).

Dr. Oyewale Tomori, then with the University of Ibadan in Nigeria, has made numerous contributions to Lassa fever research (Tomori et al. 1987, 1988, 1999; Tomori and Fabiyi 1979; Smith et al. 1979; Fabiyi et al. 1979) This includes the important observation in 1979, since confirmed in numerous populations, that many Nigerians have hemorrhagic fever antibodies in Nigerian populations (Fabiyi et al. 1979). Tomori, who is the Head of the Nigerian National Academy of Sciences, is still impacting Lassa fever research today while demanding excellence in research with grace and wit.

3.2 The 1980s

The 1980s are notable for the contributions of several extraordinary teams working on both sides of the Atlantic (Fig. 3). The team from CDC, including Drs. Joseph McCormick, Susan Fisher-Hoch, Patricia Webb, Karl Johnson, and others made numerous contributions on clinical, pathologic, virologic, epidemiologic, and therapeutic aspects of Lassa fever (Walker et al. 1982b; Trappier et al. 1993; McCormick et al. 1992; McCormick 1986). They described effective therapy with oral and intravenous ribavirin (McCormick et al. 1986), the importance of Lassa fever in children (Webb et al. 1986), and a plasma inhibitor of platelet aggregation in Lassa fever patients (Roberts et al. 1989; Cummins et al. 1989). The CDC team also performed a case-control study of *Mastomys natalensis* and humans in Lassa virus-infected households in Sierra Leone (Keenlyside et al. 1983) and a prospective study of maternal and fetal outcome in acute Lassa fever infection during pregnancy (Price et al. 1988). Aspartate transaminase levels less than 150 IU/L were shown to be a prognostic indicator of fatal Lassa fever (McCormick et al. 1987a; Johnson et al. 1987). McCormick and colleagues also confirmed that inapparent LASV infections occurred (McCormick et al. 1987b) and placed a focus on improving patient care and prevention (Fisher-Hoch et al. 1995; Fisher-Hoch et al. 1985). The CDC group developed NHP models of Lassa fever (Fisher-Hoch et al. 1987) and studied platelet function and fibrinogen in LASV-infected monkeys (Walker et al. 1982a; Lange et al. 1985).

Dr. Peter B. Jahrling and his team, who were based at United States Army Medical Research Institute of Infectious Diseases (USAMRIID), developed animal models of Lassa fever, including infections of guinea pigs, rhesus macaques, and cynomolgus macaques. Harkening back to the treatment afforded Casels, Jahrling with Dr. Clarence J. (CJ) Peters, investigated passive antibody alone and in combination with ribavirin as treatments for Lassa fever (Frame et al. 1984). In NHPs, they showed that strain-specific neutralizing antibody was important for the success

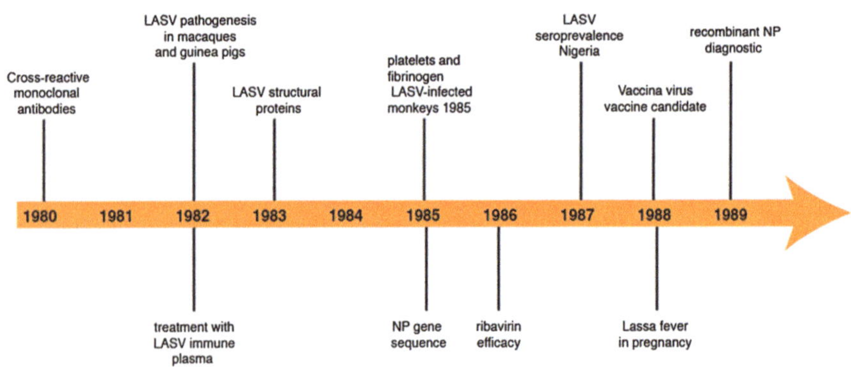

Fig. 3 Key events and scientific publications regarding Lassa fever during the 1980s

of passive antibody therapy. With John Frame of SIM and Division of Tropical Medicine, School of Public Health, College of Physicians and Surgeons, Columbia University, Jahrling also conducted a number of clinical and virological studies of Lassa fever in Liberia. Jahrling was the first Director of the National Institutes of Health (NIH) Integrated Research Facility (IRF), which like USAMRIID is on the campus of Fort Detrick (Jahrling et al. 2014).

The 1980s are also notable for contributions in the area of LASV molecular biology of Dr. J. Christopher S. Clegg and colleagues from the Special Pathogens Reference Laboratory at Porton Down in the UK. Clegg and his team described the structural proteins of LASV and sequenced and expressed the nucleocapsid gene (Clegg et al. 1986; Clegg and Lloyd 1983). They were the first to demonstrate that Lassa fever could be diagnosed using recombinant nucleocapsid antigen (Lukashevich et al. 1993; Barber et al. 1990; Lloyd et al. 1989). The team also constructed a vaccinia virus recombinant expressing LASV nucleocapsid protein and demonstrated protection against fatal Lassa fever in guinea pigs (Clegg and Lloyd 1987). A similar construct and results were reported by the CDC group, who also tested vaccinia recombinants expressing LASV glycoprotein (Morrison et al. 1989; Fisher-Hoch et al. 1989; Auperin et al. 1988).

Dr. Igor Lukashevich, now at the University of Louisville, has also sustained excellence in Lassa fever research over many decades. In the 1980s, while working at the Byelorussian Research Institute of Epidemiology and Microbiology in the United Soviet Socialist Republic (USSR) Lukashevich studied the replication of LASV in cell culture (Lukashevich et al. 1982, 1983, 1999) documented inhibition of arenavirus reproduction in cell cultures by ribavirin (Petkevich et al. 1981), demonstrated LASV lethality for inbred mice (Lukashevich 1985) and characterized LASV monoclonal antibodies. Many of Dr. Lukashevich's early studies were published in Russian. Other significance Lassa research of the 1980s includes the observation by Dr. Michael J. Buchmeier that monoclonal antibodies to lymphocytic choriomeningitis virus (LCMV) react with LASV and other pathogenic arenaviruses (Buchmeier et al. 1981; Buchmeier et al. 1980).

3.3 *The 1990s*

Several international programs continued Lassa fever research into the 1990s (Fig. 4). Sensorineural deafness was confirmed as a common sequelae of Lassa fever (Rybak 1990; Cummins et al. 1990; Cummins 1992) and the use of polymerase chain reaction (PCR) for Lassa fever diagnosis was explored (Trappier et al. 1993; Lunkenheimer et al. 1990). However, momentum for Lassa fever research was blunted by a violent civil conflict sometimes referred to as the Blood Diamonds

War that started in 1991 and continued until 2002.[1] The West African conflict cost at least 50,000 lives with local people suffering killings, mutilation, and other atrocities, mainly due to brutal warfare waged by rebel group, the Revolutionary United Front (RUF) (Campbell 2002). RUF occupied the diamond mining regions in the Lassa fever endemic regions in Eastern Sierra Leone to get access to diamonds that they traded for arms and funding to support their actions. The violence forced suspension of CDC involvement in Sierra Leone. However, admission to the Kenema Government Hospital (KGH) Lassa ward, led by Dr. Aniru Conteh, a Sierra Leonean physician continued throughout the civil conflict (Wright 2004). Lassa research continued elsewhere albeit at a slower pace. Sequencing studies were performed on the LASV L gene by Lukashevich and an international team (Lukashevich et al. 1997), and Lassa fever became further established as a significant imported tropical virus infection (Schmitz et al. 1996; Johnson and Monath 1990).

3.4 The 2000s

While their Lassa program in Sierra Leone was shut down during the Blood Diamonds War, CDC deployed Drs. Austin H. Demby and Daniel G. Bausch to neighboring Guinea. Data were collected intermittently between early 1996 and late 1999. Several significant publications from this work in Guinea appeared in the 2000s, including an investigation of the epidemiology of Lassa fever (Bausch 2001), and the distribution and prevalence of LASV in small mammals (Fig. 5) (Demby 2001). Demby, Bausch, and colleagues also performed an important comparison of enzyme-linked immunosorbent assays (ELISA), indirect fluorescent-antibody test (IFA), and virus isolation for diagnosis (Bausch et al. 2000).

The 2000s featured numerous breakthroughs that were critical for understanding Lassa virology (Fig. 5). Drs. Michael Bowen, Stuart Nichol, and others at CDC published a landmark paper on the genetic diversity of LASV (Bowen et al. 2000). Dr. William R. Gallaher published a seminal paper confirming that viral fusion proteins fit a limited number of structural scaffold and suggested the possibility that arenavirus and filovirus glycoproteins are diverged from a common ancestor (Gallaher et al. 2001). Dr. Michael B. A. Oldstone lead the group that discovered that α-dystroglycan is the receptor for Old World arenaviruses including LASV, as well as some New World arenaviruses (Spiropoulou et al. 2002; Smelt et al. 2001; Kunz et al. 2001, 2003, 2004a, b, 2005; Cao et al. 1998) Oldstone's essential work is under-represented in this review because he focused much of his work on LCMV, the prototypic arenavirus (Kunz et al. 2004a; Cao et al. 1998; Oldstone and Dixon 1970; Whitton et al. 1988; Welsh and Oldstone 1977; Welsh et al. 1977;

[1]After the "blood diamond" conflict: Lassa fever in Sierra Leone. Clin Infect Dis 59:iii–iv (2014).

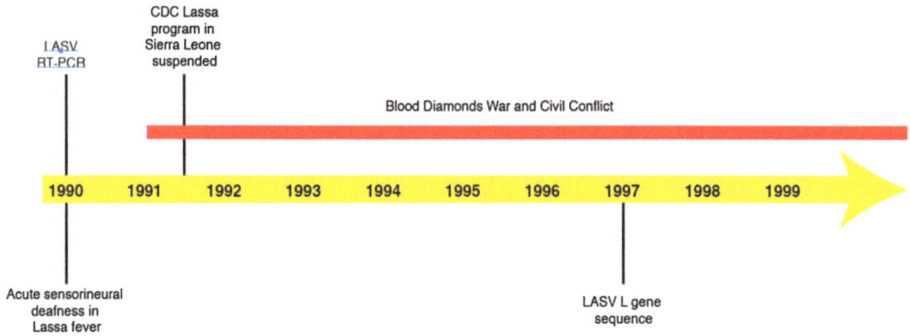

Fig. 4 Key events and scientific publications regarding Lassa fever during the 1990s

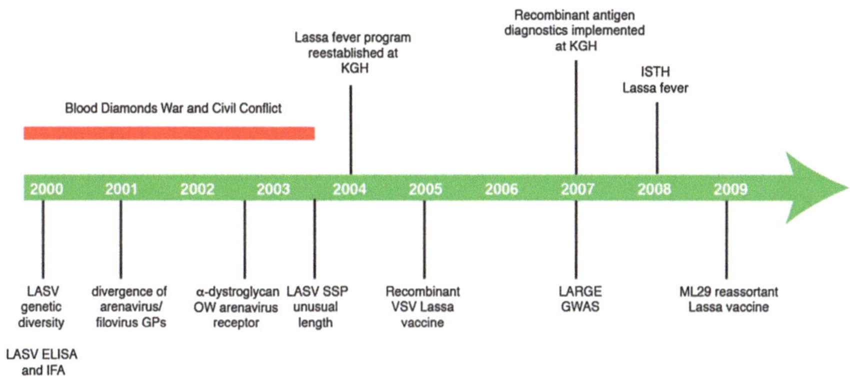

Fig. 5 Key events and scientific publications regarding Lassa fever during the 1920s

Valsamakis et al. 1986; Tishon et al. 1993, 1988; Teng et al. 1996; Sullivan et al. 2011; Southern et al. 1987; Salvato et al. 1991; Riviere et al. 1985, 1986; Oldstone et al. 1973, 1985, 1990; Oldstone and Dixon 1972; Oldstone et al. 1980; Lee et al. 2000; Klavinskis and Oldstone 1989; Ahmed et al. 1984) It was also shown that the signal peptide of the LASV glycoprotein exhibits an unusual length (Eichler et al. 2003). An additional development of note in the 2000s includes the demonstration in NHPs of the safety, immunogenicity, and efficacy of a Mopeia virus-based Lassa reassortant vaccine ML29 (Lukashevich et al. 2008).

Work in the 2000s also identified that role the zinc protein (Z) in arenavirus budding (Strecker et al. 2003). Curiously, newly described arenaviruses, the hartmaniviruses in snakes and the antennaviruses of fish, do not encode a Z protein (Garry and Garry 2019). These studies on Z have been complimented by work done on New World arenaviruses by Dr. Jack H. Nunberg and team on Junin virus (JUNV) (York and Nunberg 2007). The work of Nunberg and his collaborators on the stable signal peptide of JUNV, entry and entry inhibitors have been major contributions to knowledge about arenaviruses (York et al. 2004, 2008, 2010; York and Nunberg 2006, 2018) Other key studies of New World arenaviruses include, but are not limited to, development of a successful JUNV vaccine (Candid-1) and convalescent plasma as a treatment for JUNV infection by Dr. Delia Enria (Enria and Maiztegui 1994; Enria et al. 2008; Enria and Barrera Oro 2002; Enria et al. 2010) and structural studies on Machupo virus glycoprotein by Dr. Stephen C. Harrison (Mahmutovic et al. 2015; Abraham et al. 2010).

In 2007, Dr. Pardis C. Sabeti and colleagues at Harvard University and the Broad Institute of Harvard and MIT used long-range haplotype methods to identify human alleles that have undergone recent selection (Sabeti et al. 2007). Two populations in Nigeria exhibited signatures of selection in the gene for the LARGE protein, a glycosylase that post-translationally modifies α-dystroglycan, the cellular receptor for LASV. *DMD*, which encodes a cytosolic adaptor protein that binds to α-dystroglycan and is critical for its function, also appears to be under selection.

Following the cessation of hostilities in the West African civil conflict, a consortium of Lassa fever researchers, in close collaboration with the Sierra Leone MoHS, began rebuilding the clinical and laboratory infrastructure at KGH beginning in 2003 (Khan et al. 2008). Tragically, Dr. Conteh contracted Lassa fever and died of the Lassa fever in 2004 (Wright 2004). A then-recent graduate from the medical school in Freetown, Dr. Sheik Humarr Khan became the new director of the KGH Lassa fever program (Khan et al. 2008). The first priority for Lassa research at KGH was the development of improved diagnostic assays. Dr. Luis M. Branco, now at Zalgen Labs, developed bacterial-based systems for expression and purification of recombinant LASV proteins of immunological relevance (Branco et al. 2008). The 2000s also saw the establishment of a partnership between Sabeti's group and investigators at the Irrua Specialist Teaching Hospital (ISTH), which houses Nigeria's leading clinical and laboratory Lassa research center (Sullivan et al. 2020; Siddle et al. 2018; Okokhere et al. 2018; Okogbenin 2019; Akhuemokhan et al. 2017; Ajogbasile et al. 2020) An additional partnership at ISTH included Dr. Stephan Günther's group at the Bernhard Nocht Institute (BNI) of Tropical Medicine in Hamburg Germany (Vieth et al. 2004; Schmitz et al. 2002; Omilabu et al. 2005) Dr. Elizabeth Fichet-Calvet, who is now with the BNI group, began her important investigations on *Mastomys* and other LASV rodent hosts in the 2000s (Lecompte et al. 2006; Kerneis et al. 2009; Fichet-Calvet et al. 2007; Fichet-Calvet et al. 2008).

3.5 *The 2010s*

Lassa fever research on the decade of the 2010s was impacted by the largest outbreak of Ebola to date in the Lassa fever zone of West Africa (Fig. 6) (Hartnett et al. 2015). The West African Ebola outbreak was officially declared on March 23, 2014. Augustine Goba, director of the KGH Lassa laboratory, diagnosed the first

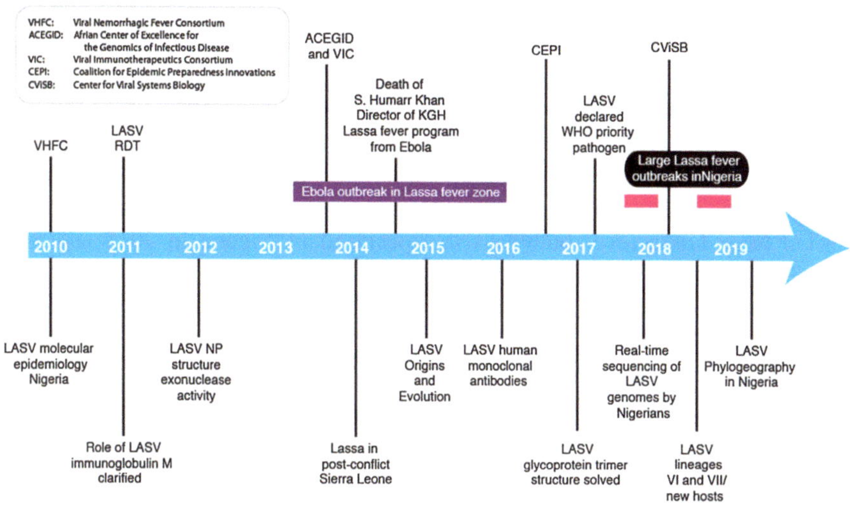

Fig. 6 Key events and scientific publications regarding Lassa fever during the 1980s

documented Ebola in Sierra Leone on May 25, 2014 (Goba et al. 2016). The majority of subsequent Ebola cases in West Africa can be traced back to a single transmission chain that includes this case (Gire et al. 2014). Thereafter, KGH received and cared for numbers of Ebola patients that quickly overwhelmed the capacity for safe management. Dr. Khan and head nurse Mbalu Fonnie were among numerous healthcare workers at KGH that contracted Ebola while administering care to infected patients and lost their lives (Bausch et al. 2014). A total of 28,616 Ebola cases with 11,310 deaths were reported in Guinea, Liberia, and Sierra Leone (Hartnett et al. 2015). An additional 36 cases and 15 deaths occurred in other countries (Hartnett et al. 2015).

The Viral Hemorrhagic Fever Consortium (VHFC), a partnership of international research institutions investigating Lassa, Ebola, and other biothreat viruses, was established in 2010. The VHFC compiled and analyzed the most complete sets of clinical and laboratory data available at the time for Lassa fever (Shaffer et al. 2014) as well as Ebola (Schieffelin et al. 2014). Recombinant antigen immunoassays for the diagnosis of Lassa fever (Boisen et al. 2018) and Ebola (Cross et al. 2016a; Boisen et al. 2016) were also developed. Sabeti and her team from Harvard/Broad developed a robust RNA-seq method for generating complete de novo assemblies of LASV and EBOV genomes in clinical and biological samples (Matranga et al. 2014). While in Sabeti's lab Dr. Kristian G. Andersen, now at Scripps Research, elucidated the spread of LASV across West Africa by genomic sequencing (Andersen et al. 2015). This migration was accompanied by changes in LASV genome abundance, fatality rates, codon adaptation, and translational efficiency. Later, Andersen founded the Center for Viral Systems Biology (CViSB; pronounced "SEE-VIZ-bee"). The mission of CViSB is to identify factors that determine outcomes from viral infections and elucidate the molecular and immunological networks that determine outcomes of human disease. Dr. Christian T. Happi of Redeemer's University established the Africa Center of Excellence for the Genomics of Infectious Diseases (ACEGID) in 2014 with support from the World Bank and NIH. ACEGID is located at the Redeemers University in Ede, Nigeria and is building the first state-of-the-art genomics laboratory on the African continent. ACEGID was well poised to respond to the protracted large Lassa fever outbreaks that occurred in Nigeria in 2018 and 2019 (Siddle 2018).

Dr. James E. Robinson of Tulane University and the VHFC team produced and characterized the first set (>120) of human monoclonal antibodies (huMAbs) to LASV (Robinson et al. 2016). A comprehensive epitope map of the Lassa virus glycoproteins was derived, including mapping the location of the major neutralizing, non-neutralizing, and linear epitopes. Drs. Thomas W. Geisbert, Chad E. Mire, and Robert W. Cross of the University of Texas Medical Branch (UTMB) in Galveston examined the ability of these antibodies to protect against fatal Lassa fever in guinea pig and nonhuman primate models (Mire et al. 2017; Cross et al. 2016a, b). The UTMB team showed that a cocktail of three LASV huMAbs each directed to a distinct epitope on GPC could reverse late symptomatic Lassa fever in *cynomolgus macaques* even with delaying treatment for more than a week post-infection (Hastie et al. 2019).

Drs. Erica Ollmann Saphire and Kathryn M. Hastie, formerly of Scripps Research and both now at the La Jolla Institute of Immunology, provided X-ray crystallographic structures of the major structural proteins of members of the arenavirus family. They solved the structure of LASV NP revealing an unexpected exonuclease activity (Hastie et al. 2011a, b, 2012a, b) and the first complete prefusion structure of the LASV glycoprotein complex (GPC) in complex with a huMAb (Hastie et al. 2017, 2019; Hastie and Saphire 2018). In 2014, Saphire established the Viral Hemorrhagic Fever Immunotherapeutic Consortium, or VIC, to develop antibody therapeutics against emerging viruses. The VIC represents a fieldwide collaboration of laboratories from around the world united to understand features that lead to antibody-mediated protection and ideal treatments for viral infections.

VHFC investigators were not of course the only Lassa researchers to make important contributions in the 2010s. Information has accumulated that LASV lineages and sub-lineages (clades) cluster geographically (Bowen et al. 2000; Siddle et al. 2018; Kafetzopoulou et al. 2019). The initial isolate of LASV (Pinneo) representing lineage I (LI) has rarely been observed in recent samplings (Ehichioya et al. 2019; Siddle et al. 2018; Olayemi et al. 2016a). LII and LIII are the most common lineages in Nigeria and are found in the southern and central regions, respectively (Ehichioya et al. 2011; Oloniniyi et al. 2018). LIV LASV is present in Sierra Leone, Liberia, and Guinea (Andersen et al. 2015; Wiley et al. 2019). LASV LV is found in Mali and Ivory Coast (Manning et al. 2015; Safronetz et al. 2013). LASV isolates from a *Hylomyscus pamfi* rodent trapped in Nigeria (Olayemi et al. 2016b) and from a nosocomial outbreak in Togo (Whitmer et al. 2018) have been proposed to represent new LASV Lineages LVI and LVII, respectively. Drs. Connie S. Schmaljohn and Katherine A. Cashman of USAMRIID showed that DNA vaccine delivered by dermal electroporation fully protects cynomolgus macaques against Lassa fever (Jiang et al. 2019; Cashman et al. 2017a, b) They also presented evidence that immune-mediated systemic vasculitis is the cause of sensorineural hearing loss in LASV-infected *cynomolgus macaques* (Cashman 2018). Dr. Schmaljohn recently became Director of the IRF.

Other significant advancements were that the World Health Organization declared LASV a priority pathogen (Mehand et al. 2018). The US CDC and Biomedical Advanced Research and Development Authority (BARDA) should follow suit and designate LASV, and other arenaviruses that cause hemorrhagic fevers as a Tier 1 agents. Pathogenic coronaviruses and paramyxoviruses should also be designated as Tier 1 agents. Another highly significant development was that founding of the Coalition for Epidemic Preparedness Innovations (CEPI), which has prioritized the accelerated development of a Lassa fever vaccine (Bernasconi et al. 2020; Gouglas et al. 2019). CEPI has also initiated an epidemiological study in five West African countries as part of their strategy to facilitate vaccine development efforts.[2]

4 Summary

The scientific literature on Lassa fever continues to expand in support of the more intense efforts against this important public health threat. Already in the decade of the 2020s near-atomic resolution structures of Lassa and Machupo virus RNA polymerases were reported (Peng et al. 2020). These structures display a similar overall architecture to polymerases of other RNA viruses, but including an arenavirus-specific domain that regulates the polymerase activity. New insights into the role of T-cell immunity against LASV have also been obtained (Sullivan et al. 2020). The next decade of research will produce important advances in the control and treatment of Lassa fever.

References

Abraham J, Corbett KD, Farzan M, Choe H, Harrison SC (2010) Structural basis for receptor recognition by New World hemorrhagic fever arenaviruses. Nat Struct Mol Biol 17:438–444. https://doi.org/10.1038/nsmb.1772

Ahmed R, Salmi A, Butler LD, Chiller JM, Oldstone MB (1984) Selection of genetic variants of lymphocytic choriomeningitis virus in spleens of persistently infected mice. Role in suppression of cytotoxic T lymphocyte response and viral persistence. J Exp Med 160:521–540

---

[2]Preparing for Lassa vaccine trials with targeted epidemiology studies, https://cepi.net/get_involved/cfps/ (2019).

Ajogbasile FV et al (2020) Real-time metagenomic analysis of undiagnosed fever cases unveils a yellow fever outbreak in Edo State, Nigeria. Sci Rep 10:3180. https://doi.org/10.1038/s41598-020-59880-w

Akhuemokhan OC et al (2017) Prevalence of lassa virus disease (LVD) in Nigerian children with fever or fever and convulsions in an endemic area. PLoS Negl Tropical Dis 11:e0005711. https://doi.org/10.1371/journal.pntd.0005711

Andersen KG et al (2015) Clinical sequencing uncovers origins and evolution of Lassa virus. Cell 162:738–750. https://doi.org/10.1016/j.cell.2015.07.020

Arnold RB, Gary GW (1977) A neutralization test survey for Lassa Fever activity in Lassa, Nigeria. Trans R Soc Trop Med Hyg 71:152–154

Auperin DD et al (1988) Construction of a recombinant vaccinia virus expressing the Lassa virus glycoprotein gene and protection of guinea pigs from a lethal Lassa virus infection. Virus Res 9:233–248

Barber GN, Clegg JC, Lloyd G (1990) Expression of the Lassa virus nucleocapsid protein in insect cells infected with a recombinant baculovirus: application to diagnostic assays for Lassa virus infection. J Gen Virol 71(Pt 1):19–28. https://doi.org/10.1099/0022-1317-71-1-19

Bausch DG et al (2000) Diagnosis and clinical virology of Lassa fever as evaluated by enzyme-linked immunosorbent assay, indirect fluorescent-antibody test, and virus isolation. J Clin Microbiol 38:2670–2677

Bausch DG et al (2001) Lassa fever in Guinea: I. Epidemiology of human disease and clinical observations. Vector Borne Zoonotic Dis 1:269–281. https://doi.org/10.1089/15303660160025903

Bausch DG et al (2014) A tribute to Sheik Humarr Khan and all the healthcare workers in West Africa who have sacrificed in the fight against Ebola virus disease: Mae we hush. Antiviral Res 111C:33–35. https://doi.org/10.1016/j.antiviral.2014.09.001

Bernasconi V et al (2020) Developing vaccines against epidemic-prone emerging infectious diseases. Bundesgesundheitsblatt, Gesundheitsforschung, Gesundheitsschutz 63:65–73. https://doi.org/10.1007/s00103-019-03061-2

Boisen ML et al (2016) Field validation of the ReEBOV antigen rapid test for point-of-care diagnosis of Ebola virus infection. J Infect Dis. https://doi.org/10.1093/infdis/jiw261

Boisen ML et al (2018) Field validation of recombinant antigen immunoassays for diagnosis of Lassa fever. Sci Rep 8:5939. https://doi.org/10.1038/s41598-018-24246-w

Bond N, Schieffelin JS, Moses LM, Bennett AJ, Bausch DG (2013) A historical look at the first reported cases of Lassa fever: IgG antibodies 40 years after acute infection. Am J Trop Med Hyg 88:241–244. https://doi.org/10.4269/ajtmh.2012.12-0466

Bowen MD et al (2000) Genetic diversity among Lassa virus strains. J Virol 74:6992–7004. https://doi.org/10.1128/jvi.74.15.6992-7004.2000

Branco LM et al (2008) Bacterial-based systems for expression and purification of recombinant Lassa virus proteins of immunological relevance. Virol J 5:74. https://doi.org/10.1186/1743-422x-5-74

Buchmeier MJ, Lewicki HA, Tomori O, Johnson KM (1980) Monoclonal antibodies to lymphocytic choriomeningitis virus react with pathogenic arenaviruses. Nature 288:486–487. https://doi.org/10.1038/288486a0

Buchmeier MJ, Lewicki HA, Tomori O, Oldstone MB (1981) Monoclonal antibodies to lymphocytic choriomeningitis and pichinde viruses: generation, characterization, and cross-reactivity with other arenaviruses. Virology 113:73–85. https://doi.org/10.1016/0042-6822(81)90137-9

Buckley SM, Casals J (1970) Lassa fever, a new virus disease of man from West Africa. 3. Isolation and characterization of the virus. Am J Trop Med Hyg 19:680–691

Buckley SM, Casals J, Downs WG (1970) Isolation and antigenic characterization of Lassa virus. Nature 227:174

Burki T (2017) CEPI: preparing for the worst. The Lancet. Infect Dis 17:265–266. https://doi.org/10.1016/s1473-3099(17)30062-2

Campbell G (2002) Blood diamond: tracing the deadly path of the world's most precious stones. Westview Press
Cao W et al (1998) Identification of alpha-dystroglycan as a receptor for lymphocytic choriomeningitis virus and Lassa fever virus. Science 282:2079–2081. https://doi.org/10.1126/science.282.5396.2079
Carey DE et al (1972) Lassa fever. Epidemiological aspects of the 1970 epidemic, Jos, Nigeria. Trans R Soc Trop Med Hyg 66:402–408
Cashman KA et al (2017a) A DNA vaccine delivered by dermal electroporation fully protects cynomolgus macaques against Lassa fever. Human Vaccines Immunotherapeutics 13:2902–2911. https://doi.org/10.1080/21645515.2017.1356500
Cashman KA et al (2017b) DNA vaccines elicit durable protective immunity against individual or simultaneous infections with Lassa and Ebola viruses in guinea pigs. Hum Vaccines Immunotherapeutics 13:3010–3019. https://doi.org/10.1080/21645515.2017.1382780
Cashman KA. et al (2018) Immune-mediated systemic vasculitis as the proposed cause of sudden-onset sensorineural hearing loss following Lassa virus exposure in cynomolgus macaques. mBio 9. https://doi.org/10.1128/mbio.01896-18
Chu M (2018). In: Barton HA, Whitaker RJ (eds). Ch. 31 Amer Soc Microbiol
Clegg JC, Lloyd G (1983) Structural and cell-associated proteins of Lassa virus. J Gen Virol 64:1127–1136. https://doi.org/10.1099/0022-1317-64-5-1127
Clegg JC, Lloyd G (1987) Vaccinia recombinant expressing Lassa-virus internal nucleocapsid protein protects guineapigs against Lassa fever. Lancet 2:186–188. https://doi.org/10.1016/s0140-6736(87)90767-7
Clegg JC, Barber GN, Chamberlain JF, Oram JD (1986) Expression of Lassa virus nucleocapsid gene fragments in bacteria. Med Microbiol Immunol 175:93–95
Cross RW et al (2016a) Analytical validation of the ReEBOV antigen rapid test for point-of-care diagnosis of Ebola virus infection. J Infect Dis. https://doi.org/10.1093/infdis/jiw293
Cross RW et al (2016b) Treatment of Lassa virus infection in outbred guinea pigs with first-in-class human monoclonal antibodies. Antiviral Res 133:218–222. https://doi.org/10.1016/j.antiviral.2016.08.012
Cross RW et al (2019) Antibody therapy for Lassa fever. Curr Opin Virol 37:97–104. https://doi.org/10.1016/j.coviro.2019.07.003
Cummins D (1992) Rats, fever and sudden deafness in Sierra Leone. Trop Doct 22:83–84
Cummins D et al (1989) A plasma inhibitor of platelet aggregation in patients with Lassa fever. Br J Haematol 72:543–548
Cummins D et al (1990) Acute sensorineural deafness in Lassa fever. JAMA 264:2093–2096
Demby AH et al (2001) Lassa fever in Guinea: II. Distribution and prevalence of Lassa virus infection in small mammals. Vector Borne Zoonotic Dis 1:283–297. https://doi.org/10.1089/153036601160025912
Eberhardt KA et al (2019) Ribavirin for the treatment of lassa fever: a systematic review and meta-analysis. Int J Infect Dis: IJID: Official Publ Int Soci Infect Dis. https://doi.org/10.1016/j.ijid.2019.07.015
ECDC (2016) Lassa fever in Nigeria, Benin, Togo, Germany and USA. Eur Centre Dis Prev Control (23 Mar 2016)
Ehichioya DU et al (2011) Current molecular epidemiology of Lassa virus in Nigeria. J Clin Microbiol 49:1157–1161. https://doi.org/10.1128/jcm.01891-10
Ehichioya DU et al (2019) Phylogeography of Lassa virus in Nigeria. J Virol. https://doi.org/10.1128/jvi.00929-19
Eichler R, Lenz O, Strecker T, Garten W (2003) Signal peptide of Lassa virus glycoprotein GP-C exhibits an unusual length. FEBS Lett 538:203–206. https://doi.org/10.1016/s0014-5793(03)00160-1
Enria DA, Barrera Oro JG (2002) Junin virus vaccines. Curr Top Microbiol Immunol 263:239–261
Enria DA, Maiztegui JI (1994) Antiviral treatment of Argentine hemorrhagic fever. Antiviral Res 23:23–31

Enria DA, Briggiler AM, Sanchez Z (2008) Treatment of Argentine hemorrhagic fever. Antiviral Res 78:132–139. https://doi.org/10.1016/j.antiviral.2007.10.010
Enria DA, Ambrosio AM, Briggiler AM, Feuillade MR, Crivelli E (2010) [Candid#1 vaccine against Argentine hemorrhagic fever produced in Argentina. Immunogenicity and safety]. Medicina 70:215–222
Fabiyi A, Tomori O (1975) Use of the complement fixation (CF) test in Lassa fever surveillance. Evidence for persistent CF antibodies. Bull World Health Organ 52, 605–608 (1975)
Fabiyi A, Tomori O, Pinneo P (1979) Lassa fever antibodies in hospital personnel in the Plateau State of Nigeria. Niger Med J 9:23–25
Fichet-Calvet E et al (2007) Fluctuation of abundance and Lassa virus prevalence in Mastomys natalensis in Guinea, West Africa. Vector Borne Zoonotic Dis 7:119–128. https://doi.org/10.1089/vbz.2006.0520
Fichet-Calvet E, Lecompte E, Koivogui L, Daffis S, ter Meulen J (2008) Reproductive characteristics of *Mastomys natalensis* and Lassa virus prevalence in Guinea, West Africa. Vector Borne Zoonotic Dis 8:41–48. https://doi.org/10.1089/vbz.2007.0118
Fisher-Hoch SP et al (1985) Safe intensive-care management of a severe case of Lassa fever with simple barrier nursing techniques. Lancet 2:1227–1229
Fisher-Hoch SP et al (1987) Physiological and immunologic disturbances associated with shock in a primate model of Lassa fever. J Infect Dis 155:465–474. https://doi.org/10.1093/infdis/155.3.465
Fisher-Hoch SP et al (1989) Protection of rhesus monkeys from fatal Lassa fever by vaccination with a recombinant vaccinia virus containing the Lassa virus glycoprotein gene. Proc Natl Acad Sci U S A 86:317–321. https://doi.org/10.1073/pnas.86.1.317
Fisher-Hoch SP et al (1995) Review of cases of nosocomial Lassa fever in Nigeria: the high price of poor medical practice. BMJ 311:857–859
Frame JD, Baldwin JM, Jr, Gocke DJ, Troup JM (1970) Lassa fever, a new virus disease of man from West Africa. I. Clinical description and pathological findings. Am J Trop Med Hyg 19:670–676
Frame JD, Casals J, Dennis EA (1979) Lassa virus antibodies in hospital personnel in western Liberia. Trans R Soc Trop Med Hyg 73:219–224. https://doi.org/10.1016/0035-9203(79)90218-9
Frame JD, Verbrugge GP, Gill RG, Pinneo L (1984) The use of Lassa fever convalescent plasma in Nigeria. Trans R Soc Trop Med Hyg 78:319–324
Fraser DW, Campbell CC, Monath TP, Goff PA, Gregg MB (1974) Lassa fever in the Eastern Province of Sierra Leone, 1970–1972. I. Epidemiologic studies. Am J Trop Med Hyg 23, 1131–1139. https://doi.org/10.4269/ajtmh.1974.23.1131
Gallaher WR, DiSimone C, Buchmeier MJ (2001) The viral transmembrane superfamily: possible divergence of Arenavirus and Filovirus glycoproteins from a common RNA virus ancestor. BMC Microbiol 1:1
Garry CE, Garry RF (2019) Proteomics computational analyses suggest that the antennavirus glycoprotein complex includes a class I viral fusion protein (alpha-penetrene) with an Internal zinc-binding domain and a stable signal peptide. Viruses 11. https://doi.org/10.3390/v11080750
Gire SK et al (2014) Genomic surveillance elucidates Ebola virus origin and transmission during the 2014 outbreak. Science 345:1369–1372. https://doi.org/10.1126/science.1259657
Goba A et al (2016) An outbreak of ebola virus disease in the lassa fever zone. J Infect Dis 214:S110–s121. https://doi.org/10.1093/infdis/jiw239
Gouglas D, Christodoulou M, Plotkin SA, Hatchett R (2019) CEPI: driving progress towards epidemic preparedness and response. Epidemiol Rev. https://doi.org/10.1093/epirev/mxz012
Hartnett JN et al (2015) Current and emerging strategies for the diagnosis, prevention and treatment of Lassa fever. Future Virol 10:559–584
Hastie KM, Saphire EO (2018) Lassa virus glycoprotein: stopping a moving target. Curr Opin Virol 31:52–58. https://doi.org/10.1016/j.coviro.2018.05.002

Hastie KM et al (2011a) Crystal structure of the Lassa virus nucleoprotein-RNA complex reveals a gating mechanism for RNA binding. Proc Natl Acad Sci U S A 108:19365–19370. https://doi.org/10.1073/pnas.1108515108
Hastie KM, Kimberlin CR, Zandonatti MA, MacRae IJ, Saphire EO (2011b) Structure of the Lassa virus nucleoprotein reveals a dsRNA-specific 3′ to 5′ exonuclease activity essential for immune suppression. Proc Natl Acad Sci U S A 108:2396–2401. https://doi.org/10.1073/pnas.1016404108
Hastie KM, Bale S, Kimberlin CR, Saphire EO (2012a) Hiding the evidence: two strategies for innate immune evasion by hemorrhagic fever viruses. Curr Opin Virol 2:151–156. https://doi.org/10.1016/j.coviro.2012.01.003
Hastie KM, King LB, Zandonatti MA, Saphire EO (2012b) Structural basis for the dsRNA specificity of the Lassa virus NP exonuclease. PLoS ONE 7:e44211. https://doi.org/10.1371/journal.pone.0044211
Hastie KM et al (2017) Structural basis for antibody-mediated neutralization of Lassa virus. Science 356:923–928. https://doi.org/10.1126/science.aam7260
Hastie KM et al (2019) Convergent structures illuminate features for germline antibody binding and pan-lassa virus neutralization. Cell 178:1004–1015.e1014. https://doi.org/10.1016/j.cell.2019.07.020
Henderson BE, Gary GW, Jr, Kissling RE, Frame JD, Carey DE (1972) Lassa fever. Virological and serological studies. Trans R Soc Trop Med Hyg 66:409–416
Jahrling PB et al (2014) The NIAID integrated research facility at Frederick, Maryland: a unique international resource to facilitate medical countermeasure development for BSL-4 pathogens. Pathog Dis 71:213–219. https://doi.org/10.1111/2049-632x.12171
Jiang J et al (2019) Immunogenicity of a protective intradermal DNA vaccine against Lassa virus in cynomolgus macaques. Human Vaccines & Immunotherapeutics 15:2066–2074. https://doi.org/10.1080/21645515.2019.1616499
Johnson KM, Monath TP (1990) Imported Lassa fever–reexamining the algorithms. N Engl J Med 323:1139–1141
Johnson KM et al (1987) Clinical virology of Lassa fever in hospitalized patients. J Infect Dis 155:456–464. https://doi.org/10.1093/infdis/155.3.456
Kafetzopoulou LE et al (2019) Metagenomic sequencing at the epicenter of the Nigeria 2018 Lassa fever outbreak. Science 363:74–77. https://doi.org/10.1126/science.aau9343
Keenlyside RA et al (1983) Case-control study of Mastomys natalensis and humans in Lassa virus-infected households in Sierra Leone. Am J Trop Med Hyg 32:829–837
Kerneis S et al (2009) Prevalence and risk factors of Lassa seropositivity in inhabitants of the forest region of Guinea: a cross-sectional study. PLoS Negl Tropical Dis 3:e548. https://doi.org/10.1371/journal.pntd.0000548
Khan SH et al (2008) New opportunities for field research on the pathogenesis and treatment of Lassa fever. Antiviral Res 78:103–115. https://doi.org/10.1016/j.antiviral.2007.11.003
Kiley MP, Lange JV, Johnson KM (1979) Protection of rhesus monkeys from Lassa virus by immunisation with closely related Arenavirus. Lancet 2:738. https://doi.org/10.1016/s0140-6736(79)90659-7
Klavinskis LS, Oldstone MB (1989) Lymphocytic choriomeningitis virus selectively alters differentiated but not housekeeping functions: block in expression of growth hormone gene is at the level of transcriptional initiation. Virology 168:232–235. https://doi.org/10.1016/0042-6822(89)90262-6
Kunz S, Sevilla N, McGavern DB, Campbell KP, Oldstone MB (2001) Molecular analysis of the interaction of LCMV with its cellular receptor [alpha]-dystroglycan. J Cell Biol 155:301–310. https://doi.org/10.1083/jcb.200104103
Kunz S, Campbell KP, Oldstone MB (2003) Alpha-dystroglycan can mediate arenavirus infection in the absence of beta-dystroglycan. Virology 316:213–220. https://doi.org/10.1016/j.virol.2003.07.002
Kunz S, Calder L, Oldstone MB (2004a) Electron microscopy of an alpha-dystroglycan fragment containing receptor sites for lymphocytic choriomeningitis virus and laminin, and use of the

receptoid body as a reagent to neutralize virus. Virology 325:207–215. https://doi.org/10.1016/j.virol.2004.04.044
Kunz S, Sevilla N, Rojek JM, Oldstone MB (2004b) Use of alternative receptors different than alpha-dystroglycan by selected isolates of lymphocytic choriomeningitis virus. Virology 325:432–445. https://doi.org/10.1016/j.virol.2004.05.009
Kunz S, Rojek JM, Perez M, Spiropoulou CF, Oldstone MB (2005) Characterization of the interaction of lassa fever virus with its cellular receptor alpha-dystroglycan. J Virol 79:5979–5987. https://doi.org/10.1128/jvi.79.10.5979-5987.2005
Lange JV et al (1985) Kinetic study of platelets and fibrinogen in Lassa virus-infected monkeys and early pathologic events in Mopeia virus-infected monkeys. Am J Trop Med Hyg 34:999–1007
Lecompte E et al (2006) *Mastomys natalensis* and Lassa fever, West Africa. Emerg Infect Dis 12:1971–1974. https://doi.org/10.3201/eid1212.060812
Lee KJ, Novella IS, Teng MN, Oldstone MB, de La Torre JC (2000) NP and L proteins of lymphocytic choriomeningitis virus (LCMV) are sufficient for efficient transcription and replication of LCMV genomic RNA analogs. J Virol 74:3470–3477. https://doi.org/10.1128/jvi.74.8.3470-3477.2000
Leifer E, Gocke DJ, Bourne H (1970) Lassa fever, a new virus disease of man from West Africa. II. Report of a laboratory-acquired infection treated with plasma from a person recently recovered from the disease. Am J Trop Med Hyg 19:677–679
Lloyd G, Barber GN, Clegg JC, Kelly P (1989) Identification of Lassa fever virus infection with recombinant nucleocapsid protein antigen. Lancet 2:1222. https://doi.org/10.1016/s0140-6736(89)91833-3
Lukashevich IS (1985) Lassa virus lethality for inbred mice. Ann Soc Belg Med Trop 65:207–209
Lukashevich IS, Vasiuchkov AD, Mar'iankova RF, Votiakov VI (1982) [Factors affecting plaque formation by Lassa virus in Vero cells]. Vopr Virusol, 57–61
Lukashevich IS, Maryankova RF, Fidarov FM (1983) Reproduction of Lassa virus in different cell cultures. Acta Virol 27:282–285
Lukashevich LS, Clegg JC, Sidibe K (1993) Lassa virus activity in Guinea: distribution of human antiviral antibody defined using enzyme-linked immunosorbent assay with recombinant antigen. J Med Virol 40:210–217
Lukashevich IS et al (1997) The Lassa fever virus L gene: nucleotide sequence, comparison, and precipitation of a predicted 250 kDa protein with monospecific antiserum. J Gen Virol 78(Pt 3):547–551. https://doi.org/10.1099/0022-1317-78-3-547
Lukashevich IS et al (1999) Lassa and Mopeia virus replication in human monocytes/macrophages and in endothelial cells: different effects on IL-8 and TNF-alpha gene expression. J Med Virol 59:552–560
Lukashevich IS et al (2008) Safety, immunogenicity, and efficacy of the ML29 reassortant vaccine for Lassa fever in small non-human primates. Vaccine 26:5246–5254. https://doi.org/10.1016/j.vaccine.2008.07.057
Lunkenheimer K, Hufert FT, Schmitz H (1990) Detection of Lassa virus RNA in specimens from patients with Lassa fever by using the polymerase chain reaction. J Clin Microbiol 28:2689–2692
Madu IG et al (2018) A potent Lassa virus antiviral targets an arenavirus virulence determinant. PLoS Pathog 14:e1007439. https://doi.org/10.1371/journal.ppat.1007439
Mahmutovic S et al (2015) Molecular basis for antibody-mediated neutralization of new world hemorrhagic fever mammarenaviruses. Cell Host Microbe 18:705–713. https://doi.org/10.1016/j.chom.2015.11.005
Manning JT, Forrester N, Paessler S (2015) Lassa virus isolates from Mali and the Ivory Coast represent an emerging fifth lineage. Front Microbiol 6:1037. https://doi.org/10.3389/fmicb.2015.01037
Matranga CB et al (2014) Enhanced methods for unbiased deep sequencing of Lassa and Ebola RNA viruses from clinical and biological samples. Genome Biol 15:519. https://doi.org/10.1186/preaccept-1698056557139770

McCormick JB (1986) Clinical, epidemiologic, and therapeutic aspects of Lassa fever. Med Microbiol Immunol (Berl) 175:153–155
McCormick JB et al (1986) Lassa fever. Effective therapy with ribavirin. N Engl J Med 314:20–26
McCormick JB, Fisher-Hoch S (1996) Level 4 Virus Hunters of the CDC (379). Turner Publications
McCormick JB et al (1987a) A case-control study of the clinical diagnosis and course of Lassa fever. J Infect Dis 155:445–455
McCormick JB, Webb PA, Krebs JW, Johnson KM, Smith ES (1987b) A prospective study of the epidemiology and ecology of Lassa fever. J Infect Dis 155:437–444. https://doi.org/10.1093/infdis/155.3.437
McCormick JB, Mitchell SW, Kiley MP, Ruo S, Fisher-Hoch SP (1992) Inactivated Lassa virus elicits a non protective immune response in rhesus monkeys. J Med Virol 37:1–7
Mehand MS, Al-Shorbaji F, Millett P, Murgue B (2018) The WHO R&D Blueprint: 2018 review of emerging infectious diseases requiring urgent research and development efforts. Antiviral Res 159:63–67. https://doi.org/10.1016/j.antiviral.2018.09.009
Mire CE et al (2017) Human-monoclonal-antibody therapy protects nonhuman primates against advanced Lassa fever. Nat Med 23:1146–1149. https://doi.org/10.1038/nm.4396
Monath TP (2019) A short history of Lassa fever: the first 10–15 years after discovery. Curr Opin Virol 37:77–83. https://doi.org/10.1016/j.coviro.2019.06.005
Monath TP et al (1973) A hospital epidemic of Lassa fever in Zorzor, Liberia, March–April 1972. Am J Trop Med Hyg 22:773–779. https://doi.org/10.4269/ajtmh.1973.22.773
Monath TP, Maher M, Casals J, Kissling RE, Cacciapuoti A (1974a) Lassa fever in the Eastern Province of Sierra Leone, 1970–1972. II. Clinical observations and virological studies on selected hospital cases. Am J Trop Med Hyg 23:1140–1149. https://doi.org/10.4269/ajtmh.1974.23.1140
Monath TP, Newhouse VF, Kemp GE, Setzer HW, Cacciapuoti A (1974b) Lassa virus isolation from *Mastomys natalensis* rodents during an epidemic in Sierra Leone. Science 185:263–265. https://doi.org/10.1126/science.185.4147.263
Morrison HG et al (1989) Protection of guinea pigs from Lassa fever by vaccinia virus recombinants expressing the nucleoprotein or the envelope glycoproteins of Lassa virus. Virology 171:179–188. https://doi.org/10.1016/0042-6822(89)90525-4
Okogbenin S et al (2019) Retrospective cohort study of lassa fever in pregnancy, Southern Nigeria. Emerg Infect Dis 25, https://doi.org/10.3201/eid2508.181299
Okokhere P et al (2018) Clinical and laboratory predictors of Lassa fever outcome in a dedicated treatment facility in Nigeria: a retrospective, observational cohort study. Lancet Infect Dis 18:684–695. https://doi.org/10.1016/s1473-3099(18)30121-x
Olayemi A et al (2016a) Arenavirus diversity and phylogeography of *Mastomys natalensis* rodents, Nigeria. Emerg Infect Dis 22:694–697. https://doi.org/10.3201/eid2204.150155
Olayemi A et al (2016b) New hosts of the Lassa virus. Sci Rep 6:25280. https://doi.org/10.1038/srep25280
Oldstone MB, Dixon FJ (1970) Tissue injury in lymphocytic choriomeningitis viral infection: virus-induced immunologically specific release of a cytotoxic factor from immune lymphoid cells. Virology 42:805–813
Oldstone MB, Dixon FJ (1972) Disease accompanying in utero viral infection. The role of maternal antibody in tissue injury after transplacental infection with lymphocytic choriomeningitis virus. J Exp Med 135:827–838. https://doi.org/10.1084/jem.135.4.827
Oldstone MB, Dixon FJ, Mitchell GF, McDevitt HO (1973) Histocompatibility-linked genetic control of disease susceptibility. Murine lymphocytic choriomeningitis virus infection. J Exp Med 137:1201–1212. https://doi.org/10.1084/jem.137.5.1201
Oldstone MB, Buchmeier MJ, Doyle MV, Tishon A (1980) Virus-induced immune complex disease: specific anti-viral antibody and C1q binding material in the circulation during persistent lymphocytic choriomeningitis virus infection. J Immunol 124:831–838

Oldstone MB et al (1985) Virus and immune responses: lymphocytic choriomeningitis virus as a prototype model of viral pathogenesis. Br Med Bull 41:70–74. https://doi.org/10.1093/oxfordjournals.bmb.a072029
Oldstone MB et al (1990) Cytotoxic T lymphocytes do not control lymphocytic choriomeningitis virus infection of BB diabetes-prone rats. J Gen Virol 71(Pt 4):785–791. https://doi.org/10.1099/0022-1317-71-4-785
Oloniniyi OK et al (2018) Genetic characterization of Lassa virus strains isolated from 2012 to 2016 in southeastern Nigeria. PLoS Negl Tropical Dis 12:e0006971. https://doi.org/10.1371/journal.pntd.0006971
Omilabu SA et al (2005) Lassa fever, Nigeria, 2003 and 2004. Emerg Infect Dis 11:1642–1644. https://doi.org/10.3201/eid1110.041343
Peng R et al (2020) Structural insight into arenavirus replication machinery. Nature 579:615–619. https://doi.org/10.1038/s41586-020-2114-2
Petkevich AS, Sabynin VM, Lukashevich IS, Galegov GA, Votiakov VI (1981) [Effect of ribovirin (virazole) on arenavirus reproduction in cell cultures]. Vopr Virusol 244–245
Pinneo L, Pinneo R (1971) Mystery virus from Lassa. Am J Nurs 71:1352–1355
Price ME, Fisher-Hoch SP, Craven RB, McCormick JB (1988) A prospective study of maternal and fetal outcome in acute Lassa fever infection during pregnancy. BMJ 297:584–587
Riviere Y, Ahmed R, Southern PJ, Buchmeier MJ, Oldstone MB (1985) Genetic mapping of lymphocytic choriomeningitis virus pathogenicity: virulence in guinea pigs is associated with the L RNA segment. J Virol 55:704–709
Riviere Y, Southern PJ, Ahmed R, Oldstone MB (1986) Biology of cloned cytotoxic T lymphocytes specific for lymphocytic choriomeningitis virus. V. Recognition is restricted to gene products encoded by the viral S RNA segment. J Immunol 136:304–307
Roberts PJ et al (1989) Plasma from patients with severe Lassa fever profoundly modulates f-met-leu-phe induced superoxide generation in neutrophils. Br J Haematol 73:152–157
Robinson JE et al (2016) Most neutralizing human monoclonal antibodies target novel epitopes requiring both Lassa virus glycoprotein subunits. Nat Commun 7:11544. https://doi.org/10.1038/ncomms11544
Rose JR (1956) A new clinical entity? Lancet 2:197
Rybak LP (1990) Deafness associated with Lassa fever. Jama 264:2119
Ryschon TW (2014) Ebola control measures and inadequate responses. Lancet 384:1181–1182. https://doi.org/10.1016/s0140-6736(14)61346-5
Sabeti PC et al (2006) Positive natural selection in the human lineage. Science 312:1614–1620. https://doi.org/10.1126/science.1124309
Sabeti PC et al (2007) Genome-wide detection and characterization of positive selection in human populations. Nature 449:913–918. https://doi.org/10.1038/nature06250
Safronetz D et al (2013) Geographic distribution and genetic characterization of Lassa virus in sub-Saharan Mali. PLoS Neg Trop Dis 7:e2582. https://doi.org/10.1371/journal.pntd.0002582
Salvato M, Borrow P, Shimomaye E, Oldstone MB (1991) Molecular basis of viral persistence: a single amino acid change in the glycoprotein of lymphocytic choriomeningitis virus is associated with suppression of the antiviral cytotoxic T-lymphocyte response and establishment of persistence. J Virol 65:1863–1869
Schieffelin JS et al (2014) Clinical illness and outcomes in patients with Ebola in sierra leone. N Engl J Med 371:2092–2100. https://doi.org/10.1056/NEJMoa1411680
Schmitz H, Emmerich P, ter Meulen J (1996) Imported tropical virus infections in Germany. Arch Virol Suppl 11:67–74
Schmitz H et al (2002) Monitoring of clinical and laboratory data in two cases of imported Lassa fever. Microbes Infect 4:43–50
Shaffer JG et al (2014) Lassa fever in post-conflict sierra leone. PLoS Negl Tropical Dis 8:e2748. https://doi.org/10.1371/journal.pntd.0002748
Siddle KJ et al (2018a) Genomic analysis of Lassa virus, 2018 Nigeria. NEJM accepted
Siddle KJ et al (2018b) Genomic analysis of lassa virus during an increase in cases in Nigeria in 2018. N Engl J Med 379:1745–1753. https://doi.org/10.1056/NEJMoa1804498

Smelt SC et al (2001) Differences in affinity of binding of lymphocytic choriomeningitis virus strains to the cellular receptor alpha-dystroglycan correlate with viral tropism and disease kinetics. J Virol 75:448–457. https://doi.org/10.1128/jvi.75.1.448-457.2001
Smith EA, Fabiyi A, Kuteyi OE, Tomori O (1979) Epidemiological aspect of the 1976 Pankshin Lassa fever outbreak. Niger Med J 9:20–22
Southern PJ et al (1987) Molecular characterization of the genomic S RNA segment from lymphocytic choriomeningitis virus. Virology 157:145–155. https://doi.org/10.1016/0042-6822(87)90323-0
Speir RW, Wood O, Liebhaber H, Buckley SM (1970) Lassa fever, a new virus disease of man from West Africa. IV. Electron microscopy of Vero cell cultures infected with Lassa virus. Am J Trop Med Hyg 19:692–694. https://doi.org/10.4269/ajtmh.1970.19.692
Spiropoulou CF, Kunz S, Rollin PE, Campbell KP, Oldstone MB (2002) New World arenavirus clade C, but not clade A and B viruses, utilizes alpha-dystroglycan as its major receptor. J Virol 76:5140–5146. https://doi.org/10.1128/jvi.76.10.5140-5146.2002
Stephen EL, Jahrling PB (1979) Experimental Lassa fever virus infection successfully treated with ribavirin. Lancet 1:268–269
Strecker T et al (2003) Lassa virus Z protein is a matrix protein and sufficient for the release of virus-like particles [corrected]. J Virol 77:10700–10705. https://doi.org/10.1128/jvi.77.19.10700-10705.2003
Sullivan BM et al (2011) Point mutation in the glycoprotein of lymphocytic choriomeningitis virus is necessary for receptor binding, dendritic cell infection, and long-term persistence. Proc Natl Acad Sci U S A 108:2969–2974. https://doi.org/10.1073/pnas.1019304108
Sullivan BM et al (2020) High crossreactivity of human T cell responses between Lassa virus lineages. PLoS Pathog 16:e1008352. https://doi.org/10.1371/journal.ppat.1008352
Teng MN, Borrow P, Oldstone MB, de la Torre JC (1996) A single amino acid change in the glycoprotein of lymphocytic choriomeningitis virus is associated with the ability to cause growth hormone deficiency syndrome. J Virol 70:8438–8443
Tishon A, Southern PJ, Oldstone MB (1988) Virus-lymphocyte interactions. II. Expression of viral sequences during the course of persistent lymphocytic choriomeningitis virus infection and their localization to the L3T4 lymphocyte subset. J Immunol 140:1280–1284
Tishon A, Eddleston M, de la Torre JC, Oldstone MB (1993) Cytotoxic T lymphocytes cleanse viral gene products from individually infected neurons and lymphocytes in mice persistently infected with lymphocytic choriomeningitis virus. Virology 197:463–467. https://doi.org/10.1006/viro.1993.1613
Tomori O, Fabiyi A (1979) Lassa fever virus: properties and characteristics. Niger Med J 9:17–19
Tomori O, Johnson KM, Kiley MP, Elliott LH (1987) Standardization of a plaque assay for Lassa virus. J Med Virol 22:77–89
Tomori O, Fabiyi A, Sorungbe A, Smith A, McCormick JB (1988) Viral hemorrhagic fever antibodies in Nigerian populations. Am J Trop Med Hyg 38:407–410. https://doi.org/10.4269/ajtmh.1988.38.407
Tomori O et al (1999) Serologic survey among hospital and health center workers during the Ebola hemorrhagic fever outbreak in Kikwit, Democratic Republic of the Congo, 1995. J Infect Dis 179(Suppl 1):S98–101. https://doi.org/10.1086/514307
Trappier SG et al (1993) Evaluation of the polymerase chain reaction for diagnosis of Lassa virus infection. Am J Trop Med Hyg 49:214–221
Troup JM, White HA, Fom AL, Carey DE (1970) An outbreak of Lassa fever on the Jos plateau, Nigeria, in January-February 1970. A preliminary report. Am J Trop Med Hyg 19:695–696
Valsamakis V et al (1986) Dissecting the molecular anatomy of persistent infection with lymphocytic choriomeningitis virus. Med Microbiol Immunol 175:97–99
Vieth S, Torda AE, Asper M, Schmitz H, Gunther S (2004) Sequence analysis of L RNA of Lassa virus. Virology 318:153–168. https://doi.org/10.1016/j.virol.2003.09.009
Walker DH et al (1982a) Experimental infection of rhesus monkeys with Lassa virus and a closely related arenavirus, Mozambique virus. J Infect Dis 146:360–368

Walker DH et al (1982b) Pathologic and virologic study of fatal Lassa fever in man. Am J Pathol 107:349–356
Watts GM (2013) Lily Lyman Pinneo. The Lancet 380:1552
Webb PA et al (1986) Lassa fever in children in Sierra Leone, West Africa. Trans R Soc Trop Med Hyg 80:577–582
Welsh RM, Oldstone MB (1977) Inhibition of immunologic injury of cultured cells infected with lymphocytic choriomeningitis virus: role of defective interfering virus in regulating viral antigenic expression. J Exp Med 145:1449–1468. https://doi.org/10.1084/jem.145.6.1449
Welsh RM, Lampert PW, Oldstone MB (1977) Prevention of virus-induced cerebellar diseases by defective-interfering lymphocytic choriomeningitis virus. J Infect Dis 136:391–399. https://doi.org/10.1093/infdis/136.3.391
White HA (1972) Lassa fever. A study of 23 hospital cases. Trans R Soc Trop Med Hyg 66:390–401
Whitmer SLM et al (2018) New Lineage of Lassa Virus, Togo, 2016. Emerg Infect Dis 24:599–602. https://doi.org/10.3201/eid2403.171905
Whitton JL, Southern PJ, Oldstone MB (1988) Analyses of the cytotoxic T lymphocyte responses to glycoprotein and nucleoprotein components of lymphocytic choriomeningitis virus. Virology 162:321–327. https://doi.org/10.1016/0042-6822(88)90471-0
WHO (2017) Lassa Fever—Benin, Togo and Burkina Faso. Disease outbreak news, 10 Mar 2017
Wiley MR et al (2019) Lassa virus circulating in Liberia: a retrospective genomic characterisation. The Lancet. Infect Dis. https://doi.org/10.1016/s1473-3099(19)30486-4
Wright P (2004) Aniru Conteh. Lancet 363:1831
Wulff H, Johnson KM (1979) Immunoglobulin M and G responses measured by immunofluorescence in patients with Lassa or Marburg virus infections. Bull World Health Organ 57:631–635
Yadouleton A et al (2019) Lassa virus in Pygmy Mice, Benin, West Africa, 2016–2017. Emerg Infect Dis 25 https://doi.org/10.3201/eid2510.180523
York J, Nunberg JH (2006) Role of the stable signal peptide of Junin arenavirus envelope glycoprotein in pH-dependent membrane fusion. J Virol 80:7775–7780. https://doi.org/10.1128/jvi.00642-06
York J, Nunberg JH (2007) A novel zinc-binding domain is essential for formation of the functional Junin virus envelope glycoprotein complex. J Virol 81:13385–13391. https://doi.org/10.1128/jvi.01785-07
York J, Nunberg JH (2018) A cell-cell fusion assay to assess arenavirus envelope glycoprotein membrane-fusion activity. Methods Mol Biol 1604:157–167. https://doi.org/10.1007/978-1-4939-6981-4_10
York J, Romanowski V, Lu M, Nunberg JH (2004) The signal peptide of the Junin arenavirus envelope glycoprotein is myristoylated and forms an essential subunit of the mature G1–G2 complex. J Virol 78:10783–10792. https://doi.org/10.1128/jvi.78.19.10783-10792.2004
York J, Dai D, Amberg SM, Nunberg JH (2008) pH-induced activation of arenavirus membrane fusion is antagonized by small-molecule inhibitors. J Virol 82:10932–10939. https://doi.org/10.1128/jvi.01140-08
York J et al (2010) An antibody directed against the fusion peptide of Junin virus envelope glycoprotein GPC inhibits pH-induced membrane fusion. J Virol 84:6119–6129. https://doi.org/10.1128/jvi.02700-09

Lassa Virus Genetics

Raphaëlle Klitting, Samar B. Mehta, Judith U. Oguzie, Paul E. Oluniyi, Matthias G. Pauthner, Katherine J. Siddle, Kristian G. Andersen, Christian T. Happi, and Pardis C. Sabeti

Contents

R. Klitting · M. G. Pauthner · K. G. Andersen (✉)
Department of Immunology and Microbiology, The Scripps Research Institute, La Jolla, CA, USA
e-mail: kga1978@gmail.com

S. B. Mehta · K. J. Siddle · P. C. Sabeti (✉)
Broad Institute of MIT and Harvard, Cambridge, MA, USA
e-mail: pardis@broadinstitute.org

S. B. Mehta
Division of Infectious Diseases, Beth Israel Deaconess Medical Center, Boston, MA, USA

J. U. Oguzie · P. E. Oluniyi · C. T. Happi
African Center of Excellence for Genomics of Infectious Diseases (ACEGID), Redeemer's University, Ede, Osun State, Nigeria

Department of Biological Sciences, Faculty of Natural Sciences, Redeemers University, Ede, Osun State, Nigeria

P. C. Sabeti
Center for Systems Biology, Department of Organismic and Evolutionary Biology, Harvard University, Cambridge, MA, USA

Department of Immunology and Infectious Diseases, Harvard TH Chan School of Public Health, Boston, MA, USA

Howard Hughes Medical Institute, Chevy Chase, MD, USA

Current Topics in Microbiology and Immunology (2023) 440: 23–65
https://doi.org/10.1007/82_2020_212

Published Online: 17 May 2020

Abstract In a pattern repeated across a range of ecological niches, arenaviruses have evolved a compact four-gene genome to orchestrate a complex life cycle in a narrow range of susceptible hosts. A number of mammalian arenaviruses cross-infect humans, often causing a life-threatening viral hemorrhagic fever. Among this group of geographically bound zoonoses, Lassa virus has evolved a unique niche that leads to significant and sustained human morbidity and mortality. As a biosafety level 4 pathogen, direct study of the pathogenesis of Lassa virus is limited by the sparse availability, high operating costs, and technical restrictions of the high-level biocontainment laboratories required for safe experimentation. In this chapter, we introduce the relationship between genome structure and the life cycle of Lassa virus and outline reverse genetic approaches used to probe and describe functional elements of the Lassa virus genome. We then review the tools used to obtain viral genomic sequences used for phylogeny and molecular diagnostics, before shifting to a population perspective to assess the contributions of phylogenetic analysis in understanding the evolution and ecology of Lassa virus in West Africa. We finally consider the future outlook and clinical applications for genetic study of Lassa virus.

1 Introduction

Lassa virus is a segmented, enveloped RNA virus in the *Arenaviridae* family, divided into genera with distinct natural hosts: mammals (*Mammarenavirus*), snakes (*Reptarenavirus* and *Hartmanivirus*), and fish (*Antennavirus*) (Radoshitzky et al. 2015; Shi et al. 2018; Hepojoki et al. 2018). The genus *Mammarenavirus* includes 35 virus species recognized by the International Committee on Taxonomy of Viruses (ICTV) as of 2018 (Maes et al. 2018), including the initial prototype lymphocytic choriomeningitis mammarenavirus (LCMV), most of which have a rodent species identified as the principal reservoir host (Forni et al. 2018). *Mammarenavirus* is further subdivided into Old and New World complexes based on genetic, geographic, and epidemiologic characteristics (Wolff et al. 1978; Charrel et al. 2002; King et al. 2012; Pontremoli et al. 2017; Brisse and Ly 2019). Figure 1 shows a phylogeny for *Mammarenavirus* species inferred from amino acid

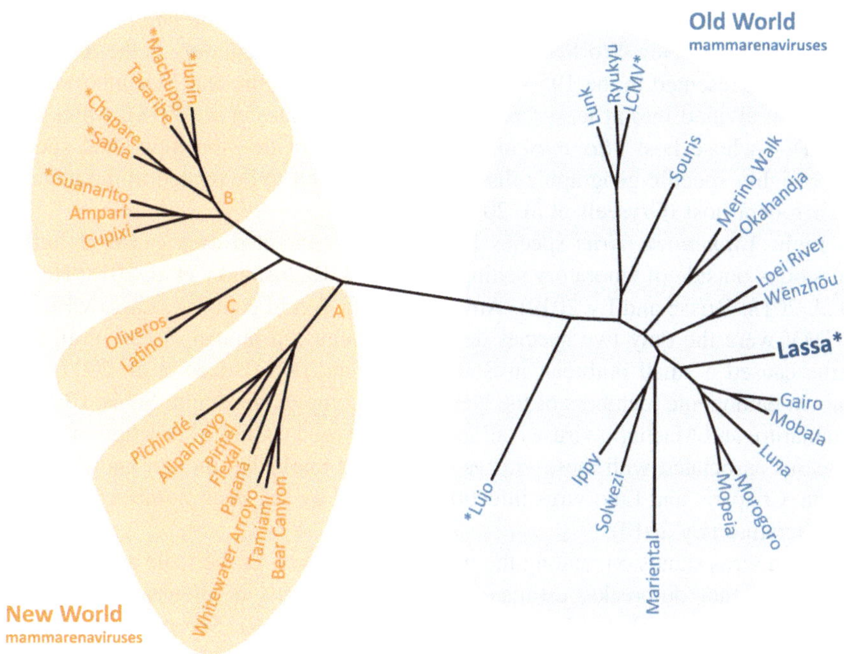

Fig. 1 Phylogenetic tree for the Mammarenavirus genus, including all members of the genus recognized by ICTV as of 2018. L gene amino acid sequences annotated in RefSeq (except Solwezi virus, where the L gene was manually marked) were aligned using MAFFT 7.388 and the tree was subsequently constructed with PhyML 3.3. The Old World complex is outlined in blue, and the New World complex, with clades A, B, and C, is outlined in orange. Viruses documented to cause natural infection in humans are marked with (*). Taxonomy here and elsewhere in this chapter uses the "Virus Name" listed by ICTV (Maes et al. 2018); in the following cases, these differ from the newer "Species Name": Junín virus (Species Name: Argentinian mammarenavirus), Sabiá virus (Brazilian mammarenavirus), Pichindé virus (Cali mammarenavirus), Paraná virus (Paraguayan), Amaparí virus (Serra do Navio mammarenavirus), and Wēnzhōu virus (Wenzhou mammarenavirus)

sequences of the RNA-dependent RNA polymerase (the L polymerase); phylogenies based on other arenavirus protein and nucleic acid sequences show the same main topology (Briese et al. 2009).

While the taxonomic classification of the genus *Mammarenavirus* in the Arenaviridae family phylogeny has been revised over the last decade (Radoshitzky et al. 2015; Maes et al. 2018, 2019), the distinction between an immunologically distinguishable Lassa–lymphocytic choriomeningitis serocomplex (now Old World arenaviruses) and Tacaribe serocomplex (now New World arenaviruses) is over half a century old (Rowe et al. 1970; Wolff et al. 1978; Charrel et al. 2002; Radoshitzky et al. 2015). The original nomenclature recognized the relationship between the globally prevalent LCMV, discovered in 1933, and Lassa virus, discovered in 1969; other viral species were gradually added to this complex, all discovered in Africa until 2015, when Wēnzhōu virus joined as the first Asian *Mammarenavirus* (Li et al. 2015). The New World complex contains species found only in the Americas, initially represented in the 1950–1960's by Tacaribe, Junín, and Machupo viruses, and now divided into at least three distinct clades (Maes et al. 2018). Apart from LCMV—whose host *Mus musculus* is found worldwide—the individual species have highly specific geographic distributions believed to be related to the niche of their rodent host (Gryseels et al. 2017).

Eight *Mammarenavirus* species have been documented as a cause of human infections outside of laboratory settings (Ly 2017; Radoshitzky et al. 2018; Hallam et al. 2018; Brisse and Ly 2019). Among the Old World complex, Lassa virus and LCMV were the only two species described as causing human disease until Lujo virus caused a small outbreak in South Africa in 2008 (Briese et al. 2009). The known pathogenic members of the New World complex are Junín, Sabía, Chapare, Guanarito, and Machupo viruses, all in clade B (Fig. 1). Clinical descriptions of the diseases associated with these viruses often differ markedly, though knowledge of Sabía, Chapare, and Lujo virus infection is limited by the small number of cases in the literature (Ly 2017).

Lassa virus stands out among the mammarenaviruses for the scale and regularity of its seasonal outbreaks, estimated to affect hundreds of thousands of people annually (McCormick et al. 1986; Eberhardt et al. 2019). Seropositivity against Lassa virus appears higher than predicted by case counts, suggesting that many infections result in mild or non-specific symptoms that are never formally diagnosed (McCormick and Fisher-Hoch 2002). In its most severe form, however, Lassa fever results in dysfunction of multiple organ systems and coagulopathy leading to hemorrhage, with high inpatient case fatality rates (Andersen et al. 2015). The only available antiviral treatment for Lassa fever is the nucleoside analog ribavirin, denoted an essential medicine by the World Health Organization, although its use is often limited by adverse effects (McCormick et al. 1986). There are currently no other effective medications or preventative vaccines outside of clinical trials.

2 Molecular Genetics

Deciphering the molecular genetics and biology of Lassa virus is critical to understanding its life cycle and pathology, and it can inform the development of new diagnostics and interventions. Laboratory analysis is particularly challenging for Lassa virus, which, along with Lujo virus and the pathogenic New World *Mammarenavirus* species, is a select agent, restricting study of replication-competent wild-type strains to highly specialized biosafety level 4 (BSL-4) conditions. LCMV, first isolated in the 1930's (Armstrong and Lillie 1934; Rivers and McNair Scott 1935), has often served as a basis for the study of general *Mammarenavirus* biology, including Lassa virus (Djavani et al. 1997), since experimental work on this species can occur in less stringent settings. Biology specific to Lassa virus can then either be explored in BSL-4 facilities or using reverse genetic techniques that create recombinant or non-infectious forms of Lassa virus that require lower biocontainment levels. This range of approaches has been used to study the replication cycle (Hass et al. 2004, 2006; Pinschewer et al. 2005; Albariño et al. 2011) and virus–host interactions (Wang et al. 2018a; Willard et al. 2018). Reverse genetic schemes in particular have been further refined into platforms for antiviral drug screening (Caì et al. 2018; Madu et al. 2018; Zhang et al. 2019) and development of candidate vaccines (Lukashevich and Pushko 2016; Martínez-Sobrido and de la Torre 2017; Purushotham et al. 2019).

2.1 Genome Structure and Function

An understanding of the structure of the *Mammarenavirus* genome and the functions of its gene products is fundamental to any discussion of Lassa virus genetics. As noted above, many of these insights were initially obtained using LCMV as a prototypic *Mammarenavirus*, and they are either assumed or have later been demonstrated to be true for Lassa virus. Below we provide an introduction to this literature, which has been extensively discussed in several recent reviews (Hallam et al. 2018; Loureiro et al. 2019; Brisse and Ly 2019).

2.1.1 Genome Organization

Lassa virus and other mammarenaviruses have a linear, ambisense RNA genome comprised of two single-stranded segments, known as the short ($\sim$3.4 kb) S and long ($\sim$7.2 kb) L (Lukashevich et al. 1984). Each segment has two oppositely oriented open reading frames for a total of four genes (Auperin et al. 1984; Meyer et al. 2002; Fig. 2): the S segment encodes a $\sim$63 kDa viral nucleoprotein NP and a $\sim$75 kDa glycoprotein precursor peptide GPC, while the L segment encodes a $\sim$200 kDa RNA-dependent RNA polymerase L and a $\sim$11 kDa RING finger protein Z (Hass et al. 2004). The 5′ untranslated region (UTR) of each segment is complementary to the 3′ UTR of the same segment, allowing each to form a terminal 19–33 nucleotide RNA panhandle that modulates replication and transcription (Albariño et al. 2011). The non-coding intergenic region between the two open reading frames can form a stable hairpin structure that also regulates the viral life cycle (Salvato and Shimomaye 1989; Fig. 2).

2.1.2 Nucleoprotein NP

Arenaviral genomic RNA is encapsidated by the nucleoprotein NP to form a nucleocapsid, which in turn associates with the L polymerase to form the ribonucleoprotein (RNP) particle (Young and Howard 1983; Bruns and Lehmann-Grube 1983). This core structural function of NP oligomers is a universal feature used by

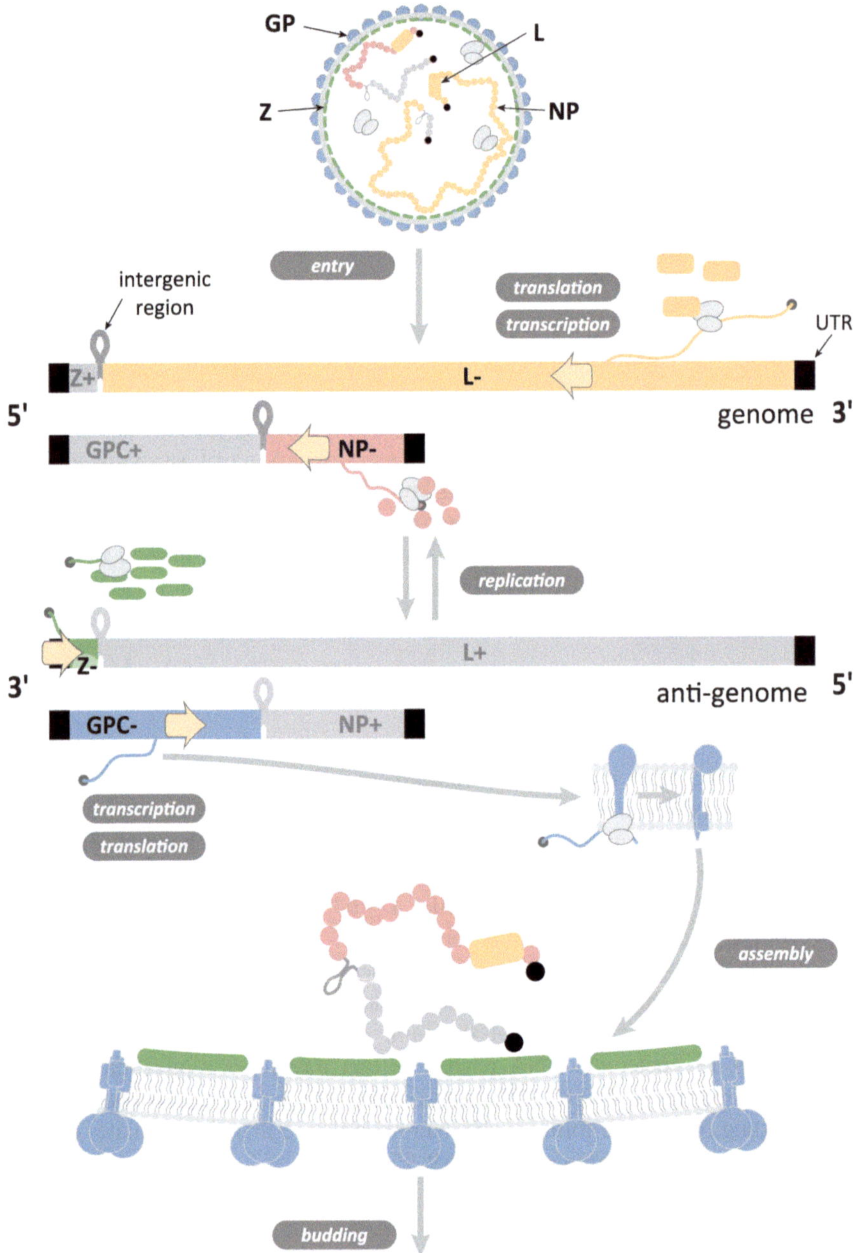
GP
L
Z
NP
entry
intergenic region
translation
transcription
UTR
Z+
L-
5'
genome
3'
GPC+
NP-
replication
Z-
L+
3'
anti-genome
5'
GPC-
NP+
transcription
translation
assembly
budding

◂**Fig. 2 Molecular organization and biology of Lassa virus**. The Lassa virion consists of genomic RNA, nucleoprotein, and polymerase associated as ribonucleoprotein particles which are in turn associated with matrix proteins and packaged with ribosomes into a lipid envelope. The viral genomic short segment carries a negative-sense nucleoprotein (NP) gene and a positive-sense glycoprotein precursor (GPC) gene, while the long segment carries negative-sense RNA-dependent RNA polymerase (L) and a positive-sense matrix (Z) genes; each segment has matching 5' and 3' untranslated regions (UTR) and an intergenic hairpin loop. Upon entry into the host cell cytoplasm, the L polymerase directly transcribes negative-sense genes from the template 3' UTR to the intergenic region, resulting in mRNAs for the L and NP genes. The L polymerase can also continue past the intergenic region to create full-length segments that act as an antigenome, allowing transcription of Z and GPC mRNAs and in turn serving as a template to replicate the original genome. The GPC peptide is post-translationally modified and targeted to the membrane, where the mature multimeric glycoprotein interacts with the Z protein and ribonucleoprotein particles to assemble and bud new virions

negative-strand RNA viruses to package, protect and control access to the viral genome (Ruigrok et al. 2011; Reguera et al. 2014), and in mammarenaviruses the resulting complex is both necessary and sufficient for viral transcription and replication (Lee et al. 2000; López et al. 2001; Hass et al. 2004) on entering the host cell cytoplasm. The arenavirus NP is known to have other functions critical to the viral life cycle, including association with the Z protein for viral assembly (reviewed in Loureiro et al. 2012) and exoribonuclease activity that suppresses the host type I interferon response (King et al. 2017; Huang et al. 2018; Iwasaki et al. 2018).

2.1.3 L Polymerase

The L gene encodes a multifunctional enzyme with RNA-dependent RNA polymerase, host-cap-snatching endonuclease, and regulatory domains (Vieth et al. 2004; Brunotte et al. 2011; Vogel et al. 2019). Packaged in the viral RNP, this polymerase can transcribe mRNA from the two negative-sense genes, L and NP, upon fusion with the host cell. Messenger RNA synthesis begins at the 3′ end of each genomic segment and is believed to use a prime-and-realign mechanism, where one to five extra, non-templated residues are created at the mRNA 5′ terminus by allowing the first few transcribed nucleotides to slip slightly in the template's 3′ direction before transcription continues along the gene (Raju et al. 1990; Garcin and Kolakofsky 1992; Meyer and Southern 1993; Polyak et al. 1995; Marq et al. 2010, 2011). The L polymerase and NP protein together modify these non-templated 5′ residues by adding a 7-methylguanylate cap (Raju et al. 1990; Lelke et al. 2010; Qi et al. 2010) which interferes with the host ability to recognize viral mRNA and may improve mRNA stability and translation efficiency (Both et al. 1975; Furuichi et al. 1977; Foscaldi et al. 2017).

Transcription of negative-sense genes by the L polymerase continues until the non-coding intergenic region. The resulting non 3′ poly-adenylated mRNAs vary in final length, suggesting that transcription is terminated by a structural interaction between the polymerase and the intergenic hairpin rather than a specific stop sequence (Meyer and Southern 1993; Meyer et al. 2002; Pinschewer et al. 2005).

The interaction between viral polymerase and intergenic region appears to be regulated such that transcription events either generate mRNA in a transcriptive mode or generate a full-length antigenome by continuing through the predicted hairpin in a replicative mode (Ferron et al. 2017), although the control signals that modulate between these two modes are not yet understood. Once full-length copies are available, transcriptional activity of the L polymerase on the antigenomic template results in GPC and Z gene mRNAs, while another replicative cycle results in a copy of the original viral genome.

The need for an antigenome is a direct result of the arenavirus ambisense coding scheme, and it naturally suggests a mechanism for temporal sequencing of transcription. Acute infection of a cell leads to an early phase of viral transcription where L and NP mRNAs dominate, since those open reading frames are immediately available for transcription. Once the L polymerase produces antigenomes, the infection enters a late phase where GPC and Z gene products are produced and the genome itself can be replicated for packaging into a new virion (Fuller-Pace and Southern 1988; Meyer et al. 2002; King et al. 2018).

2.1.4 Glycoprotein Precursor GPC

After the Lassa virus pre-glycoprotein polyprotein GP complex (GPC) is transcribed and translated, the resulting polypeptide is targeted to the rough endoplasmic reticulum by its N-terminal signal sequence. This signal sequence is then cleaved by the host protein SPase to produce a stable signal peptide (SSP), which stays associated with the nascent GP and is required for maturation of the GP (Eichler et al. 2003). The remainder of the GPC polypeptide is glycosylated and cleaved into GP1 and GP2 subunits, forming a SSP, GP1, and GP2 heterotrimer that is dispatched to the cellular plasma membrane (Lenz et al. 2001; Rojek et al. 2008). The now-extracellular GP1 and SSP remain noncovalently associated with the transmembrane GP2, and this tripartite complex in turn trimerizes to form a class I viral fusion glycoprotein spike which in turn mediates receptor recognition and viral entry into a host cell (Eschli et al. 2006; Burri et al. 2012; Hastie and Saphire 2018).

Host cell infection begins when the Lassa virus surface glycoprotein's GP1 extracellular spike binds to host α-dystroglycan (α-DG), an extracellular protein that links the extracellular matrix to the cell's actin cytoskeleton via the transmembrane ß-dystroglycan. Post-translational processing of α-DG involves extensive glycosylation (reviewed in Godfrey et al. 2007), and the action of the glycosyltransferase LARGE on α-DG has specifically been shown to be critical for binding of LCMV, Lassa, and other arenaviruses (Kunz et al. 2005a; Hara et al. 2011). Binding of a virion to the cell surface triggers endocytosis (Cao et al. 1998) and delivers virions to the endosomal system. The acidic pH of late endosomes causes GP1 to switch targets from α-DG to lysosomal-associated membrane protein 1, triggering structural changes that allow the viral particle to fuse with the containing lysosome and release its contents into the host cell cytoplasm (Li et al. 2016; Pasquato et al. 2018).

2.1.5 Z Protein

The Z protein coordinates the packaging and budding of infectious particles, recruiting "endosomal sorting complex required for transcription" proteins to facilitate virus egress (Capul et al. 2011; Urata and Yasuda 2012). Despite its relatively small size, the Z protein has multiple domains, which are conserved across arenaviruses, and appear to have multiple functions. A myristoylation site at the N-terminus is required for association with the intracellular plasma membrane and subsequent budding of virions (Perez et al. 2004). A RING domain in the center of the protein chelates zinc ions and appears important for Z-NP and Z-LP interactions required to package ribonucleoproteins into the forming virion (Wilda et al. 2008; Casabona et al. 2009; Shtanko et al. 2010; Capul et al. 2011; Loureiro et al. 2011; Ortiz-Riaño et al. 2011). Other conserved sequences in the Z protein have less understood roles, although mutations in these sequences influence budding, packaging, and regulation of viral RNA synthesis to various degrees (Wang et al. 2012).

2.1.6 Persistence

Unlike its acute course in humans, Lassa virus can cause a persistent infection in its reservoir host, *Mastomys natalensis* (Walker et al. 1975; Mariën et al. 2017), with ongoing viral replication but without overt host illness. This pattern of infection has been extensively studied in LCMV as a model of a non-cytopathic virus that avoids immune clearance in its natural host (Rawls et al. 1981; Kane and Golovkina 2010; Flint et al. 2015). Mammarenavirus interactions with the host immune system are complex, utilize different mechanisms between species and between Old World and New World branches, and can depend on the age and route of initial infection (e.g. Oldstone 2007). In culture, LCMV readily establishes stable infection without causing widespread cytotoxicity. The possible protection of infected cells from superinfection with new virions, in vivo production of defective viral particles, and detailed temporal control of transcription, have all been proposed as mechanisms to avoid overwhelming cellular machinery with virus production (Rawls et al. 1981; Rojek et al. 2007a; Gaudin and Kirchhausen 2015; King et al. 2018). It is not yet understood which of these balanced immune and cellular mechanisms in natural hosts fail in incidental hosts such as humans, where the outcome of infection is either virus clearance or death, without evidence for sustained viremia.

2.2 *Reverse Genetics*

Reverse genetics approaches investigate the function of a gene by analyzing the phenotype resulting from known mutations, as opposed to the forward genetic approach of searching for the genotype corresponding to a given phenotype. The application to RNA viruses involves setting up a reverse genetics system, which is a procedure to recreate or rescue an infectious virion from its RNA or cDNA alone. One can use this system to rescue virions from mutated or even more substantially edited viral genomes, and proceed to characterize the resulting phenotypes. These systems can further be harnessed to produce crippled viruses that only perform some of the viral functions, to examine specific steps of the viral life cycle.

2.2.1 Rescue of Bona Fide Viruses

The naked, ambisense genomic RNA of a mammarenavirus is not infectious. The transcription and replication both require the activity of the L polymerase, which uses the genomic RNA associated with nucleoprotein NP as a template, which is also called the nucleocapsid (Fig. 2). Hence, for a mammarenavirus, the minimum unit of infectivity, defined as the fewest elements able to start a productive replication cycle, consists of the genomic segments S and L along with the viral proteins NP and L.

The first mammarenavirus to be rescued from cDNA was LCMV in 2006, using sets of four or five plasmids (Flatz et al. 2006; Sánchez and de la Torre 2006), with similar strategies subsequently applied to Junín (Emonet et al. 2011) and Machupo viruses (Patterson et al. 2014). The core of this approach was to provide the NP and L genes *in trans* (using expression plasmids), and to direct the synthesis of the virus genomic segments with either a T7 RNA polymerase (RP) or a mouse pol-I promoter system (Fig. 3a; Palese 1995; Neumann et al. 1999). The NP protein interacts with the transcribed genome segments to form the nucleocapsid. After this artificial encapsidation, the L polymerase interacts with the nucleocapsid to create the RNP complex, which is necessary and sufficient to start a full viral life cycle (Hallam et al. 2018). Infectious viral particles rescued using this procedure exhibit in vitro and in vivo replicative fitness similar to that of *bona fide* viruses (Flatz et al. 2006; Sánchez and de la Torre 2006). This strategy was subsequently refined into a simpler two-plasmid system (Fig. 3b), first applied to Junín (Albariño et al. 2009), and then Pichindé (Lan et al. 2009), Lassa (Albariño et al. 2011; Carnec et al. 2011) and Lujo viruses (Bergeron et al. 2012). The two-plasmid system drives the expression of the antigenomic segments S and L using a T7 RP promoter. The L and NP proteins are directly translated from the antigenomic segments in sufficient amounts to initiate genome replication, triggering the viral life cycle (Albariño et al. 2009).

With the two approaches described above, authentic Lassa virus strains can be obtained entirely de novo from synthetic or cloned DNA constructs rather than having to isolate live virus. Another example for a BSL-4 pathogen is the Ituri variant from the North Kivu Ebola outbreak that began in 2018, and was rescued using a similar approach (McMullan et al. 2019). The efficiency of diagnostic assays, antiviral drugs, and neutralizing antibodies treatments could thus be evaluated on circulating viral strains, while the outbreak was ongoing, without requiring that the virus first be isolated and transported. The ability to generate authentic viruses de novo also set new paths to investigations of the biology of Lassa virus, allowing researchers to edit non-coding parts of the genome and to investigate their

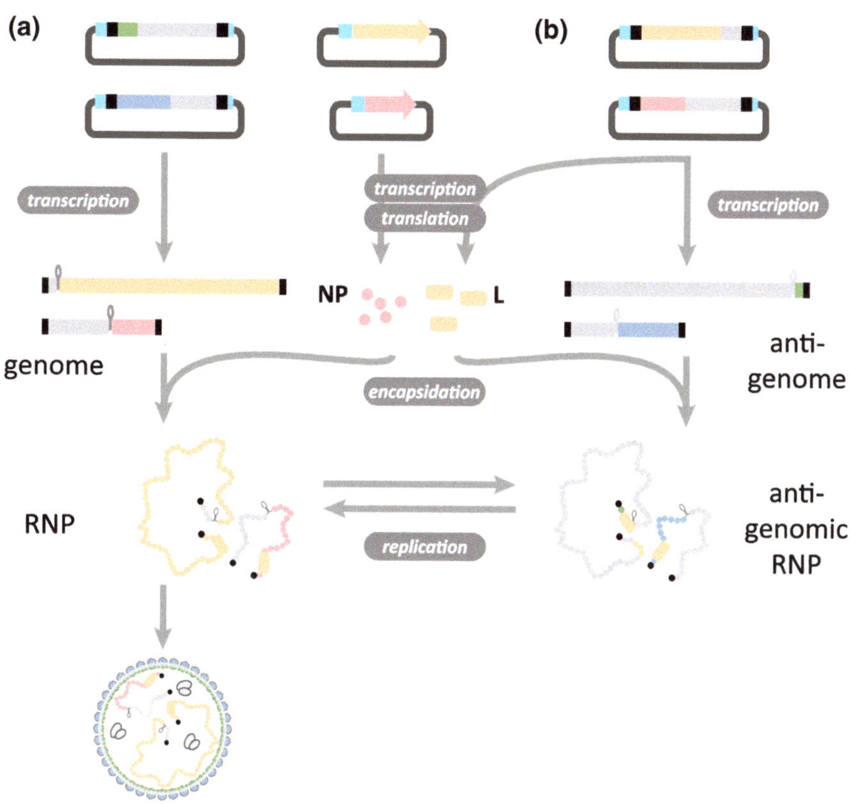

Fig. 3 Reverse genetics systems for the study of the Lassa virus replication cycle. Modified from (Beitzel et al. 2019). (**a**) Four-plasmid and (**b**) two-plasmid systems are depicted above. In (**a**), two plasmids support synthesis of the genomic RNAs under the control of a promoter system (T7 or pol-I, in light blue), while a terminator sequence (pol-I or hepatitis delta virus ribozyme, also in light blue) ensures the generation of a precise 3′ end for the transcribed RNA. The two other plasmids support the expression of the NP (in red) and L (in dark blue) proteins under the control of a T7 or a pol-II promoter (in light blue). Genomic RNAs are subsequently encapsidated by NP, and the L protein interacts to form functional RNPs and trigger the replication cycle. With the two-plasmid system in (**b**), the plasmids support the synthesis of the antigenomic RNAs and NP and L proteins. NP and L proteins join the segments to form antigenomic RNPs that can replicate to yield genomic RNP's and trigger a normal viral replication cycle

importance for in vivo virulence (Albariño et al. 2011), to generate mutants of Lassa virus nucleoprotein for functional studies (Carnec et al. 2011) and to produce tagged versions of the Z protein, to detect Lassa virus-infected cells by flow cytometry (Schaeffer et al. 2019).

2.2.2 Recombinant Viruses

Live Lassa virus must be handled under BSL-4 conditions, which results in technical constraints on experiments and higher costs. Systems have been set up to circumvent these constraints using recombinant viruses by starting with a less pathogenic virus such as LCMV and substituting one of its genes with a compatible Lassa virus gene. Since the combined genes must be compatible for the recombinant viruses to efficiently replicate, the majority of recombinant Lassa virus constructs are based on a mammarenavirus backbone. This approach has the advantage of generating a viral entity that is biologically close to the wild-type virus, but can in most cases be handled safely in lower biosafety level settings.

The procedure used to generate these recombinant viruses is analogous to the one used to rescue *bona fide* Lassa virus, with L and NP proteins provided *in trans* and genome segments expressed using a T7 RNA polymerase or a pol-I system. In this case, however, the genomic segments include genes or segments from two different viral species, such as Lassa virus GPC with LCMV NP, L and Z proteins (Lee et al. 2013) or Lassa virus S segment with Mopeia virus L segment (Lukashevich et al. 2005), although not all combinations result in a replicative fitness sufficient for use in experiments. The Lassa virus GPC in a LCMV backbone has been one of the most popular recombinant systems to date, having served in screenings for the antiviral activity of Lassa virus glycoprotein inhibitors (Urata et al. 2011) and for numerous studies of Lassa virus entry (Pasqual et al. 2011; Goncalves et al. 2013; Oppliger et al. 2016; Fedeli et al. 2018).

Recombinant virus technology has also been extensively used in the development of live-attenuated vaccine candidates. The Mopeia/Lassa virus reassortant ML29 (Lukashevich et al. 2005; Johnson et al. 2019) in particular appears to be safe and immunogenic in non-human primate models (Lukashevich et al. 2008; Zapata et al. 2013). Viral backbones other than *Mammarenaviruses*, such as the vesicular stomatitis virus have also been used to express Lassa virus antigens, and proved to be protective in vivo against strains from Liberia, Mali, and Nigeria (Geisbert et al. 2005; Safronetz et al. 2015).

Finally, recombinant techniques can add foreign genes of interest (GOI) to a wild-type virus for use in high-throughput drug screens or live-attenuated vaccine platforms. This technique was implemented in LCMV by Emonet *et al.* (2009) using a three-segment system, including two S segments and one L segment. In each S segment, one viral ORF was replaced by a reporter gene, ensuring that only virions including the two different S segments can replicate. This approach yielded LCMV virions comparable in shape and size to the wild-type virus, with similar in vitro replicative fitness (Emonet et al. 2009; Popkin et al. 2011). Of note, some tri-segmented recombinant *Mammarenaviruses* can have impaired growth properties in vitro (Dhanwani et al. 2015) or in vivo (Emonet et al. 2009). This system has been used for LCMV, Junín, Pichindé and Machupo viruses (Emonet et al. 2009, 2011; Dhanwani et al. 2015; Zaza et al. 2018), but not yet established for Lassa virus.

Genes of interest can also be integrated directly into a viral genomic segment, although this strategy can result in genomic instability and risks losing expression of the integrated gene over time (Emonet et al. 2009). Recombinant Lassa virus (rLASV) genomes with a cleavable GFP (Caì et al. 2018) or ZsGreen protein (Welch et al. 2016) inserted next to the NP gene have been created. The rLASV expressing ZsGreen was not assessed for stability, but rLASV-GFP did maintain a stable expression of GFP over 10 in vitro passages (Caì et al. 2018). A drug discovery screen and an antibody neutralization assay were both developed based on rLASV-GFP.

2.2.3 Defective Viral Particles

Non-infectious Lassa virus surrogates also permit study outside of BSL-4 laboratories. Considering the ensemble of procedures available to date, it is now possible to deconstruct the viral life cycle into discrete functions and steps to provide a detailed characterization of Lassa virus biology, from the investigation of viral replication and gene expression, to the virus assembly and budding.

Replicon or minigenome systems support the replication and expression of shortened versions of the genomic RNA without allowing production of infectious particles. The principle is to use a viral genome that contains the cis-acting factors, including the intergenic loop and both 3′ and 5′ UTRs, necessary for the expression and the replication of the genome, but in which one or more viral genes have been replaced by reporter genes. This minimal version of the genome can be expressed, replicated, and packaged, but since it does not code for functional viral proteins, these proteins must be provided *in trans*.

Minigenomes have been developed for LCMV (Lee et al. 2000), Lassa virus (Hass et al. 2004; Welch et al. 2016), Tacaribe virus (López et al. 2001), Lujo virus (Bergeron et al. 2012) and Machupo virus (Patterson et al. 2014). This approach has helped to characterize the Lassa virus promoter sequences (Hass et al. 2006), the matrix function of the Z protein (Capul et al. 2011), and has been used to screen drugs for antiviral activity (Hass et al. 2004; Welch et al. 2016).

Lassa virus pseudotyped particles constitute another convenient BSL-2 surrogate to study the Lassa virus entry step. Since the envelope glycoprotein GP drives receptor binding and fusion with the host cell, a particle coated with Lassa virus GP can mimic the entry of the virus into its target cell (Reignier et al. 2006). Pseudotyped particles are produced by coexpressing the envelope protein of interest (here of Lassa virus) together with the genome of a supporting virus in which the native envelope protein has been deleted or impaired. Different platforms have been used to produce Lassa virus GP pseudotyped particles, including vesicular stomatitis virus, HIV, murine leukemia virus, and LCMV (Cosset et al. 2009; Whitt 2010; Rodrigo et al. 2011; Robinson et al. 2016), providing insights into Lassa virus entry and virus-receptor interactions (Kunz et al. 2005b; Reignier et al. 2006; Rojek et al. 2007b, 2012; Shimojima et al. 2012; Cohen-Dvashi et al. 2016). Lassa pseudotypes have also been valuable in neutralization assays (Robinson et al. 2016), high-throughput screening of entry inhibitors (Lee et al. 2008; Wang et al. 2018b) and vaccine development (Branco et al. 2010; Li et al. 2017). Also based on this strategy, a single-cycle LASV replicon vaccine candidate has been implemented, in which the supporting virus is Lassa virus itself (Kainulainen et al. 2018).

Finally, Lassa virus-like particles can be formed by expressing the Z and GPC peptides alone or in combination. Both proteins self-assemble into a structure that resembles the shape of the wild-type virus but are not infectious (Strecker et al. 2003; Schlie et al. 2010a). These particles have been used to study the structure (Strecker et al. 2003; Li et al. 2016) and composition (Schlie et al. 2010b; Watanabe et al. 2018) of Z and GP protein complexes, to identify inhibitors of viral entry (Lee et al. 2008; Mohr et al. 2015) and to dissect the process of viral budding (Urata et al. 2012). Recently, a vaccinia virus-based vaccine candidate expressing Lassa GPC and Z has been described, but elicited only low levels of antibodies in mice (Salvato et al. 2019).

3 Genome Sequencing and Diagnostics

From the isolation of Lassa virus in 1970 (Buckley and Casals 1970) to the beginning of the year 2000, complete or nearly complete genome sequence data was available in public databases for only three viral strains: the Nigerian strains Pinneo and GA391 and the Sierra Leonean strain Josiah (Bowen et al. 2000). From this limited information, the first PCR assays were designed, but their performance varied widely between strains from different countries due to the diversity between geographic sites (Lunkenheimer et al. 1990; Trappier et al. 1993). As more genome-level sequencing data became available, a deeper understanding of this diverse species has offered insight into viral evolution and enabled the development of better molecular diagnostics. Ongoing technological development to increase the speed, sensitivity, and cost-effectiveness of both Lassa virus sequencing and clinical diagnostics promises to improve detection, surveillance, and broaden our understanding of Lassa virus.

3.1 Sequencing

Lassa virus was initially recognized to have significant sequence diversity based on short sequences (Ter Meulen et al. 1998) and T1-oligonucleotide mapping (Gonzalez et al. 1988). Since 2000, more than 1000 genomic sequences have become available (Fig. 4), enabled by the development of massively parallel sequencing technologies. Rapid innovation in this field in recent years aims to balance increased sensitivity, depth of coverage and throughput with minimizing costs and other barriers to adoption of sequencing methods.

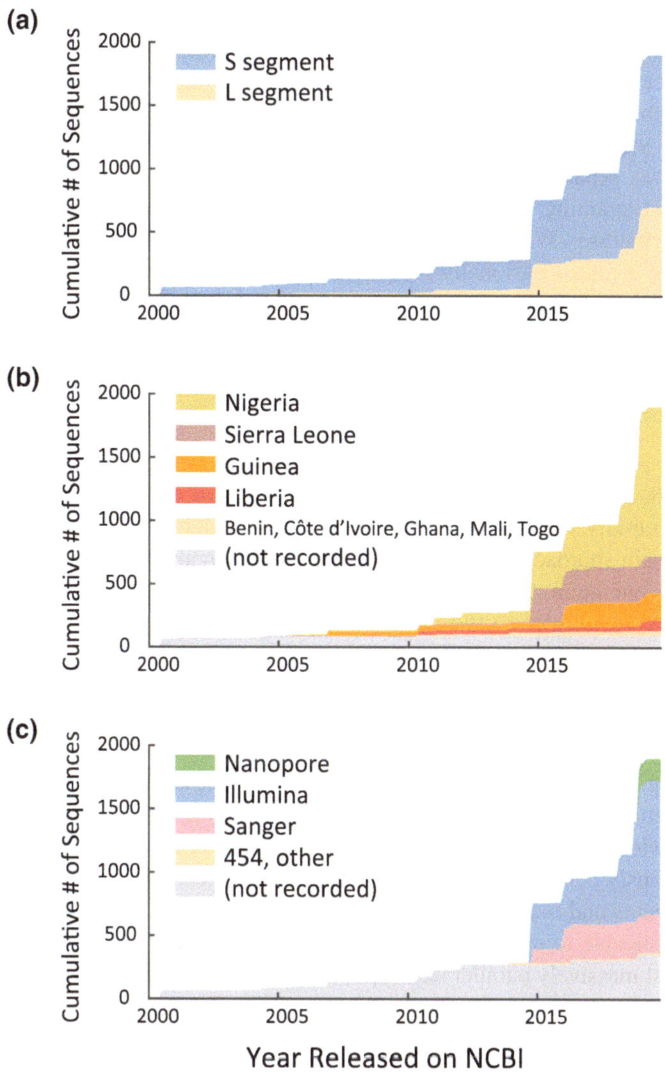

Fig. 4 Lassa virus nucleotide sequences in NCBI GenBank. (**a**) Cumulative number of samples in GenBank over time, divided by genomic segment. (**b**) Cumulative number of samples in GenBank, divided by country of origin. (**c**) Cumulative number of samples in GenBank, divided by sequencing technology. Counts are as of September 2019; in all cases "(not recorded)" implies that the Comment field in GenBank did not include the relevant information

3.1.1 Sanger Sequencing

Early Lassa virus sequences were generated by Sanger, or chain-terminator sequencing, the principal sequencing technology for over thirty years (Shendure et al. 2017). Sequencing was labor-intensive, requiring either cultured or cloned virus, or the amplification of a specific genomic fragment by RT-PCR. The S segment of the Josiah strain was first cloned and sequenced by Auperin and McCormick (1989), while Bowen et al. (Bowen et al. 2000) developed an alternative method for amplification and cloning of the full S segment RNA of arenaviruses, yielding the complete gene sequences for NP, GP1, and GP2. The L segment was separately cloned and sequenced (Djavani et al. 1997; Lukashevich et al. 1997), granting insight into the function of the L protein or RNA-dependent RNA polymerase. While these approaches have been largely replaced, Sanger sequencing is still occasionally used today for sequencing of smaller amplicons or cultured isolates (Oloniniyi et al. 2018; Ehichioya et al. 2019).

3.1.2 Unbiased Sequencing

Massively parallel sequencing has largely superseded Sanger sequencing methods, dramatically increasing throughput (Shendure et al. 2017). These technologies have advanced rapidly with a proliferation of platforms, which can be broadly separated into two main approaches; (i) short-read sequencing methods, sometimes referred to as Next Generation Sequencing and (ii) real-time, single-molecule sequencing methods. Technologies in the former group (e.g., Illumina and Ion Torrent) typically generate large numbers of reads in the range of a few hundred bases, while the latter class of technologies (e.g., PacBio and Oxford Nanopore) generate fewer, longer reads (up to a megabase). When implemented in a sequence-independent manner—often termed metagenomics—these technologies allow a shared, unbiased sequencing pipeline to detect genetic material, and potentially assemble genomes, for, theoretically, any species. For pathogen genomics, this alleviates the need for prior knowledge or suspicion of the target organism. A metagenomic approach can simultaneously perform rapid identification of pathogens, genomic characterization of viral strains and mutations, and identification of coinfections (Gardy and Loman 2018), applicable to outbreaks or ongoing surveillance.

The first massively parallel techniques to be broadly adopted used sequencing by synthesis of very large numbers of short DNA fragments in parallel, with a fluorescence-based readout. The majority of Lassa virus genome sequences have been generated using such methods, in particular using Illumina sequencing platforms (Fig. 4c). The preparation of a typical sequencing library from RNA extracted from a clinical sample involves RNA-dependent synthesis of cDNA with random priming, followed by fragmentation with addition of adaptors for sequencing and barcodes for multiplexing by pooling multiple samples (Matranga et al. 2016). Illumina sequencing typically produces reads of length 100–300 nucleotides, compared to 500–800 nucleotide reads from Sanger sequencing, and the large number of resulting reads has spurred development of computational methods for species identification (Ye et al. 2019) and genome assembly (Ayling et al. 2019). These technologies have been widely adopted, and they have enabled rapid in-country sequencing of Lassa samples in Nigeria (Andersen et al. 2015; Siddle et al. 2018; Kafetzopoulou et al. 2019).

More recently, real-time, single-molecule sequencing has also been applied to Lassa virus sequencing using Oxford Nanopore Technologies' minION device (Kafetzopoulou et al. 2019). Based on detecting minute changes in current as individual DNA strands pass through nanometer-scale pores, the minION is capable of generating much longer sequencing reads, with a size and technological requirements that are markedly smaller and lighter than alternative massively parallel technologies. This approach does have a higher error rate than sequencing by synthesis, though this may be mitigated by sequencing to higher depth, use of various error-correction tools (Rang et al. 2018) or combining with short-read sequencing data. The portability of the minION allows it to be rapidly deployed to the site of an outbreak, exemplified by Kafetzopoulou et al. (2019) at the Irrua Specialist Teaching Hospital in Edo State, Nigeria, decreasing delays and sample degradation due to transport.

3.1.3 Targeted Sequencing

Clinical and field samples present a sequencing challenge, since in many instances the viral load is too low, or the host nucleic acid background too high, to obtain sufficient data using the unbiased approach described above. In these cases, target enrichment or PCR-amplification prior to massively parallel sequencing can increase sensitivity, decrease cost, and provide greater depth of sequencing (Houldcroft et al. 2017).

Target enrichment, also sometimes referred to as hybrid capture, uses DNA or RNA oligonucleotide probes that hybridize to complementary sequences of interest. Following hybridization, unbound DNA fragments are washed away, depleting host and other non-targeted sequences. Bound fragments are then eluted and amplified before sequencing using standard massively parallel techniques. For Lassa virus, Matranga et al. (2014) designed a hybrid capture technique using a pool of oligonucleotide probes, designed to cover the known diversity of Lassa virus, to enrich viral nucleic acid in unbiased sequencing libraries. More recently, Metsky *et al.* (2019) extended this approach with a computational tool capable of designing minimal probe sets that fully cover the sequence diversity of target species. They then used this tool to develop a probe set that covers sequence diversity among all viral species known to infect humans and used this to successfully obtain complete or partial Lassa virus genome assemblies from very low titer clinical samples. Such multi-virus probe sets—also developed by others (Briese et al. 2015; Wylie et al. 2015)—are especially interesting as it can increase sensitivity without restricting detection to a single virus.

Sensitivity can also be improved by initially amplifying RNA using target-specific primers in RT-PCR (Houldcroft et al. 2017). This can either be accomplished by amplification of the entire target genome followed by fragmentation to shorten amplified DNA before sequencing or by parallel amplification of overlapping short pieces of the target genome with a multiplexed library of tiled primers. The tiled approach has been successfully applied to other RNA viruses including Zika virus (Quick et al. 2017), Chikungunya virus (Naveca et al. 2019), and yellow fever virus (Faria et al. 2018). Lassa virus' sequence diversity, however, presents a challenge for the design of specific primers to amplify full genomic segments, limiting primer site options for a tiled approach or requiring a highly multiplexed primer pool. In contrast, the relative conservation of the 3′ and 5′ UTR of both S and L segments (Fig. 2) makes a whole genome amplification approach appealing, though the ability to recover sequence information from degraded RNA may be compromised. Beyond improving overall sensitivity, this approach can achieve deep coverage of the genome from only hundreds of thousands of reads, rather than the tens of millions required to achieve the equivalent depth with unbiased approaches, potentially enabling affordable detection of genetic diversity within the viral population infecting a single patient.

3.2 *Nucleic Acid Diagnostics*

Lassa fever can be mistaken for other infections that also circulate in its endemic range, particularly early in the disease course. Since both false negatives and false positives have serious implications for treatment efficacy, patient isolation, and infection control, rapid, accurate, and accessible laboratory diagnostics for Lassa virus are crucial (Raabe and Koehler 2017). In 2018, the WHO identified Lassa fever under its R&D blueprint as a priority threat and highlights challenges of existing diagnostics (WHO 2018). Diagnosis with viral culture is difficult, slow, and hazardous (Demby et al. 1994), and detection of antibodies has limited utility for the first one to two weeks after symptom onset (Wulff and Johnson 1979; Bausch et al. 2000; Branco et al. 2011). While antigen capture may eventually prove to be the most convenient front-line rapid diagnostic test, antigen-based assays require extensive development and optimization (Boisen et al. 2018), while nucleic-acid based diagnostics can be designed and deployed rapidly (e.g., Hui et al. 2020).

3.2.1 PCR

Reverse transcription polymerase chain reaction (RT-PCR) is currently the method of choice for early detection of Lassa virus from clinical samples in Lassa endemic countries (Olschlager et al. 2010; Olschläger and Günther 2012; Salami et al. 2020). While dramatic advances in PCR technologies have substantially addressed early concerns about cost and complexity (Lunkenheimer et al. 1990), the diversity of the Lassa virus genome continues to pose a challenge.

The first RT-PCR assays to detect Lassa virus used primer designs based on the Josiah strain from Sierra Leone, but these primers showed poor sequence homology (Lunkenheimer et al. 1990) and failed to amplify (Trappier et al. 1993) isolates from Nigeria. Demby et al. (1994) subsequently designed primers that targeted the 5′

conserved UTR and GPC gene of the S segment using eight sequences in addition to the Josiah strain, including some from Nigeria and Liberia, but this assay too performed less well on strains from outside of Sierra Leone (Olschlager et al. 2010). This assay was subsequently revised using 62 available S sequences, shifting the reverse primer and including 2 degenerate bases (Olschlager et al. 2010), giving improved performance for Nigerian and Liberian strains. The inclusion of degenerate bases, while extremely useful for amplifying genetically diverse targets, broadens the melting temperature of the reaction, making reaction optimization more difficult. The diversity within amplified sequences further limits the use of probe-based reporters such as TaqMan (Olschläger and Günther 2012). Quantitative PCR for Lassa virus has thus most commonly used non-specific DNA-binding dyes such as SYBR or multiplexing strain-specific probes (Trombley et al. 2010; Nikisins et al. 2015; Boisen et al. 2018).

New RT-PCR assay designs continue to be proposed for Lassa virus detection (Raabe and Koehler 2017; Mazzola and Kelly-Cirino 2019; Takah et al. 2019). The majority of these tests target the S segment (Mazzola and Kelly-Cirino 2019), since the L segment has fewer available sequences and typically shows lower RNA levels in clinical samples (Andersen et al. 2015). Many laboratories continue to use in-house assays, with limited external quality assessments historically driven by the research community (Nikisins et al. 2015; Takah et al. 2019), although clinical diagnostic laboratories in Lassa endemic areas, notably Kenema Government Hospital in Sierra Leone and Irrua Specialist Teaching Hospital in Nigeria, have increasingly been evaluating assay designs (Trombley et al. 2010; Asogun et al. 2012; Nikisins et al. 2015; Boisen et al. 2018). Recently, the World Health Organization (WHO) has emphasized the need for a standardized and validated assay, which has been designated a priority threat (Salami et al. 2020). The current gold standard emerged from the commercial release in 2015 of an RT-PCR assay produced by Altona RealStar® though it is currently sold only for research use, with a recently updated v2.0 kit further seeking to improve accuracy by combining probe-based assays for the L and S segment.

3.2.2 Beyond PCR

Although they are currently the method of choice for Lassa virus detection across much of the endemic region, PCR-based assays present a number of challenges, in particular the need for relatively sophisticated laboratory equipment and the difficulty of designing primers that can detect genetically diverse strains. To this end, several other non-PCR-based nucleic acid amplification tests have been proposed, including techniques based on loop mediated isothermal amplification (Fukuma et al. 2011; Pemba et al. 2019), nanostring (Koehler et al. 2018), and microarrays (Das et al. 2015). Loop-mediated isothermal amplification has been of particular interest since the absence of thermal cycling simplifies infrastructure requirements, but the stringent primer design requirements have kept a single pan-Lassa virus assay out of reach, and existing tests are clade-specific (Fukuma et al. 2011; Pemba et al. 2019). Modifications to traditional PCR, such as Aldatu Biosciences' PANDAA^{TM} assay, which overcomes target diversity by incorporating site-directed mutagenesis (Duarte et al. 2017), are also in development. CRISPR-based diagnostics, such as Specific High Enzymatic Reporter Unlocking (SHERLOCK) and DNA endonuclease—targeted CRISPR trans reporter (DETECTR) (Gootenberg et al. 2017; Chen et al. 2018; Myhrvold et al. 2018), are being developed to use a lateral-flow based paper readout—similar to those used for the promising antigen-capture rapid diagnostic test ReLASV. This may become particularly valuable for Lassa viral diagnostics since the equipment and training required for existing nucleic acid amplification diagnostics limits their use outside of laboratory settings (Boisen et al. 2018). Ultimately, this diversity of technological efforts will require standardization, evaluation, and quality control for validation and eventual regulatory approval (Mazzola and Kelly-Cirino 2019).

4 Population Genomics

Advances in the sensitivity, affordability, and accessibility of sequencing technologies have led to rapid growth in the number of publicly available Lassa virus genome sequences (Bowen et al. 2000). This increase has been dominated by a small number of endemic regions, notably Nigeria, which in recent years has accounted for an increasing proportion of documented cases (Fig. 4b; ProMED-mail 2018, 2020; Shaffer et al. 2019), and Sierra Leone, as well as smaller numbers from Liberia and Guinea. In well-sampled regions, sequence data clarifies the extent and spatial structure of viral genetic diversity and the phylogenetic relationships between lineages. This description in turn enables studies of general trends at the species level, assessment of transmission patterns, and a deeper understanding of the evolutionary history of the virus. On smaller scales, viral phylogenies can further be used to investigate local circulation dynamics and to complement epidemiological data.

4.1 Species Phylodynamics

The Lassa virus species includes at least four major geographically and genetically distinct lineages, with a fifth and sixth recently proposed (Fig. 5), and the analysis of these diverse viral sequences and their relatedness using phylogenetic inference allows us to explore the dynamics of Lassa virus evolution. We first review the molecular basis for genetic diversity within the species and then explore the implications of this diversity on transmission patterns and the inferred historical spread across West Africa.

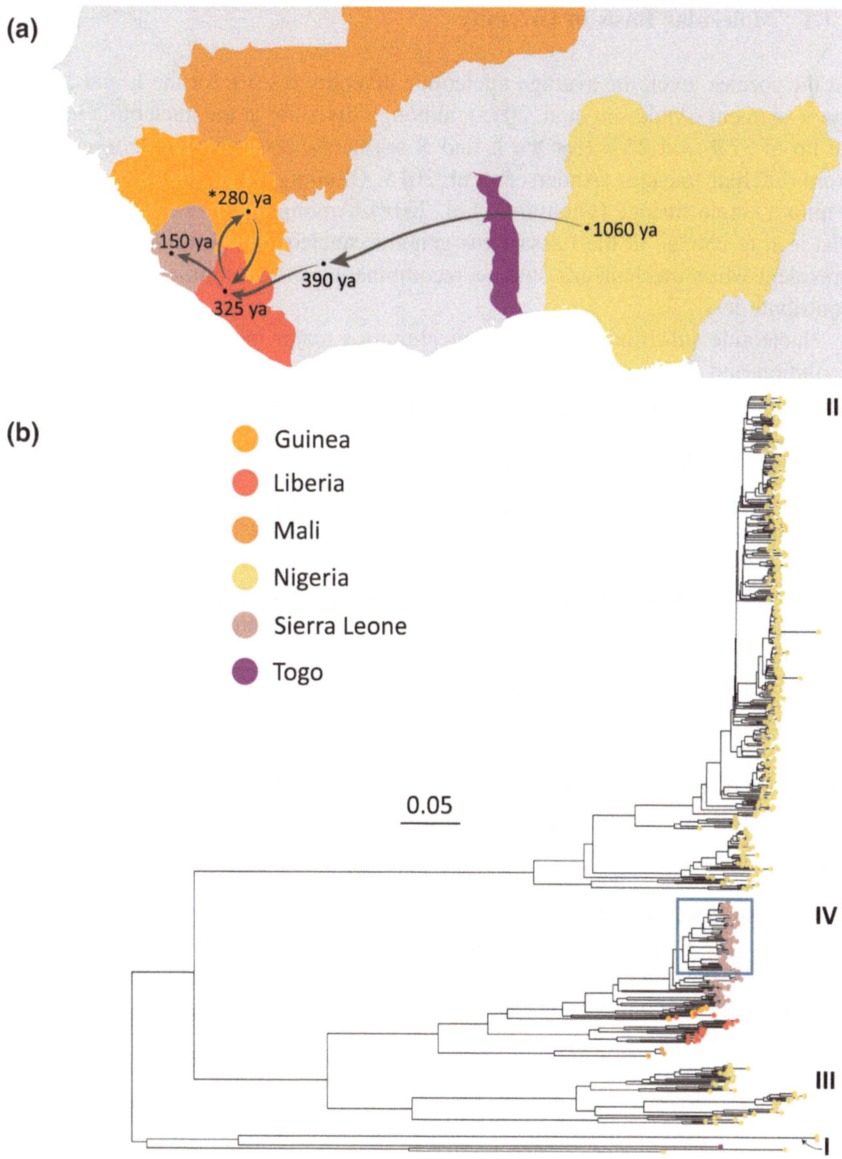

Fig. 5 Lassa virus spread. (**a**) Potential spread of Lassa virus based on L segment tMRCA's, given as median values (ya = years ago) and based on published estimates (Andersen et al. 2015; Wiley et al. 2019); for the suspected introductions of Lassa into Guinea, only the oldest tMRCA is given. (**b**) Phylogeny of Lassa virus based on 613 S segment complete ORF sequences, labeled with the current view of major lineages (adapted from Andersen et al. 2015); more recently described lineages V and VI are not shown. The blue box marks the region shown in more detail in Fig. 6. The scale bar is in units of nucleotide substitutions per site

4.1.1 Molecular Basis of Diversity

At the species level, the average nucleotide diversity is 20% for the L and 15% for the S segment (Andersen et al. 2015), although diversity at the nucleotide level can be up to 32% and 25% (for the L and S segments, respectively) between strains from different lineages (Andersen et al. 2015; Oloniniyi et al. 2018), or up to 10% within a single lineage (Oloniniyi et al. 2018). Among the molecular mechanisms that lead to change in the Lassa virus genome, nucleotide substitution seems most prevalent while mechanisms such as recombination or reassortment are thought to contribute less.

Nucleotide substitution during replication is a major molecular basis of genetic evolution and RNA viruses often generate significant genetic diversity with substitution rates on the order of 10^{-3} substitutions/site/year (Jenkins et al. 2002); the substitution rate for Lassa virus is in this range, having been estimated at $\sim 9 \times 10^{-4}$ substitutions/position/year (Andersen et al. 2015; Fichet-Calvet et al. 2016). Beyond this fundamental mechanism, segmented viruses have the potential to create reassortant viruses when more than one strain infects the same cell. Mammarenaviruses have been shown to have evidence of reassortment, and even more complex mixing of genomic segments, within- (Stenglein et al. 2015) and potentially between-species (Lukashevich 1992; Charrel and de Lamballerie 2003; Lukashevich et al. 2005), but this is not believed to be a major component in Lassa virus evolution. While Lassa viruses can reassort in vitro (Lukashevich 1992), there is currently limited evidence for intra-species reassortment in nature, with only ~1–2% of sequenced samples demonstrating patterns consistent with reassortment (Andersen et al. 2015; Whitmer et al. 2018; Kafetzopoulou et al. 2019; Wiley et al. 2019). Similarly, recombination, the exchange of sub-segment fragments between viruses in a single cell, has never been formally reported for Lassa virus, despite being observed in other arenaviruses (Charrel et al. 2001, 2002; Archer and Rico-Hesse 2002; Stenglein et al. 2015). Only a small number of Lassa strains, including the Kako, Jirandogo, and Togo isolates, show phylogenetic patterns suggestive of recombination, namely differences in clustering depending on which portion of the S segment is used for tree reconstruction (Kronmann et al. 2013; Olayemi et al. 2016; Whitmer et al. 2018).

4.1.2 Transmission Patterns

Understanding transmission networks can provide critical information to inform measures for virus control. Genomic data complements traditional epidemiological approaches, such as contact tracing, to inform our understanding of virus transmission. When a reasonable proportion of cases is sampled, phylogenies can approximate relationships between cases (Fig. 6a). They also allow to both identify general transmission trends and connect chains of highly similar sequences to suggest direct or proximal transmission to inform transmission networks (Wohl et al. 2016). Sequence information overwhelmingly supports reservoir spillover (Fig. 6b) as the

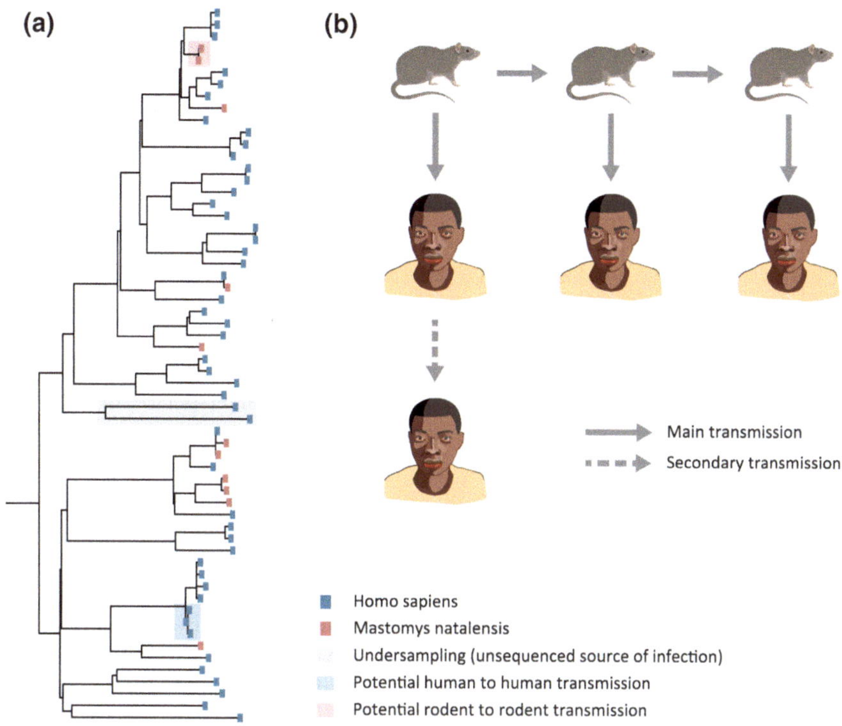

Fig. 6 Lassa virus transmission dynamics. (**a**) subtree from Lassa virus phylogenetic tree in Fig. 5. Tips are colored here by the host from which the sequence was obtained. Examples of clusters indicating potential rodent-to-rodent or human-to-human transmission are highlighted. The gray box shows long branches with no closely related viruses which have been sequenced, suggesting undersampling. (**b**) Schematic of Lassa virus transmission. The main mechanisms are depicted by solid arrows. The virus is maintained in the primary reservoir species *Mastomys natalensis* and spread from rodent-to-rodent, with most human cases resulting from rodent-to-human spillover. Human-to-human transmission, indicated by a dashed arrow, is believed to play a limited role

primary mode of Lassa virus transmission, consistent with predictions from both rodent biology and ecology. Phylogenetic analysis can further rapidly reassess these dynamics in response to unexpected changes in the pattern of reported infections (Andersen et al. 2015; Siddle et al. 2018; Kafetzopoulou et al. 2019).

Lassa virus infection can be acquired from rodent or human sources, directly through exposure to infected body fluids or indirectly through contaminated foods and surfaces (McCormick 1987; Ogbu et al. 2007). Although the earliest documented Lassa virus outbreaks featured nosocomial transmission (Carey et al. 1972; Monath et al. 1973), direct evidence for significant human-to-human transmission has since been rare and the majority of cases are believed to result from exposure to infected rodents or their excreta (McCormick and Fisher-Hoch 2002). Considering the risks of increased human-to-human transmission, ongoing monitoring of patterns of Lassa virus transmission is crucial to inform outbreak control and prevention, particularly when peaks in Lassa fever activity are observed (Siddle et al. 2018; Kafetzopoulou et al. 2019).

Genomic studies (e.g., Andersen et al. 2015) provide multiple lines of evidence to support the view that most Lassa fever cases are due to independent spillover events from a persistently infected reservoir into a largely dead-end human population (Fig. 6b), rather than a rapid chain-like spread of the virus, as observed during the West African Ebola outbreak in 2014–2016 (Gire et al. 2014). Genomic evidence for primarily rodent-to-human transmission has been detailed in three major studies (Andersen et al. 2015; Siddle et al. 2018; Kafetzopoulou et al. 2019). First, human samples tend to cluster more strongly by geography than by collection date. Second, when rodent and human sequences from the same area have been studied, sequences from the two host taxa are generally interspersed in viral phylogenies, without the distinct clusters of human samples that would suggest evolution separate from the rodent reservoir. In contrast, clusters among contemporaneous rodent sequences are observed (see, e.g., Fig. 6a), indicating rodent-to-rodent transmission consistent with the role of *M. natalensis* as the Lassa virus reservoir host. When examining the correlation between sample collection date and the genetic distance of that sample from the root of the phylogenetic tree, the sustained, directional spread exemplified by human transmission chains should appear as a strong correlation over the months-to-years time scale of Lassa fever outbreaks, referred to as a clock-like signal. Lassa virus phylogenies do not, however, show a molecular clock signal for Lassa virus, indicating that nucleotide substitutions are not accumulating in humans on these time scales and suggesting that Lassa virus is largely circulating and evolving in a well-mixed rodent reservoir, with occasional spillover into humans. Recent work showed that this trend persisted during the surge of cases reported in Nigeria during the 2017–2018 transmission season (Siddle et al. 2018; Kafetzopoulou et al. 2019).

While the topology of Lassa virus phylogenetic trees does suggest that overall human-to-human transmission is a minor contributor to the spread of Lassa fever, pairs or small clusters of samples consistent with human-to-human transmission have been identified (Andersen et al. 2015; Siddle et al. 2018; Kafetzopoulou et al. 2019). These clusters may reflect a stuttering transmission pattern in which zoonotic spillover events lead to short human transmission chains (e.g., Fig. 6b). Another possibility would be that these clusters correspond to large but undersampled transmission chains, though this is inconsistent with existing non-genomic evidence that does not suggest extensive human-to-human transmission (McCormick and Fisher-Hoch 2002). In a number of cases, epidemiologic or genomic proximity alone cannot definitively demonstrate a route of infection: both human- and rodent- derived transmission occur in similar settings, such as a single home or village, and identical or nearly identical virus isolates might equally be contracted from a common local rodent reservoir. In this context, within-host variants may provide an avenue to improve reconstruction of transmission networks from genomic data (Wohl et al. 2016). A within-host variant is defined as the subset of virions in a single host that share a point mutation; the mutations themselves are often referred to as intrahost single nucleotide variations or iSNVs. Since chronically infected *M. natalensis* harbor more within-host diversity than is seen in human cases (Andersen et al. 2015; Siddle et al. 2018; Kafetzopoulou et al. 2019), comparison of variants in an infected individual with that of potential rodent sources may help understand the origin of an infection. Further characterization of the within-host dynamics and transmission bottlenecks is needed to assess the value of this approach for Lassa virus.

4.1.3 Ancient Dispersal

Techniques that incorporate sampling date when building phylogenetic trees can estimate a time to the most recent common ancestor (tMRCA) for divergence events between extant viral lineages. These estimates can be used to investigate the development of an outbreak. They may also provide valuable information about the evolution of a virus over a broader, inter-epidemic scale to describe the long-term persistence of a virus or its spread in its reservoir host (Grubaugh et al. 2019). When timed phylogenies show a consistent spatial pattern, they can further inform the geographic routes by which a virus may have spread.

Although no strong clock-like signal is identifiable for Lassa viral sequences over time scales of months to years, molecular clock behavior is evident over longer time scales (Andersen et al. 2015). Based on this correlation, Bayesian coalescent analysis (Drummond et al. 2012) can place lower-bounds on the tMRCAs for major Lassa lineages that is, estimate the latest possible time at which these lineages separated from each other. Since these estimates depend on sampling of existing diversity, tMRCA estimates may be pushed back in time if new divergent sequences are encountered and proposed dispersal routes may change as samples from new locations become available.

Most studies estimate a tMRCA of all sequenced Lassa virus older than 1,000 years ago (Ehichioya et al. 2011; Andersen et al. 2015; Manning et al. 2015; Oloniniyi et al. 2018). In viral phylogenies spanning the known diversity of Lassa virus species, the deepest branches are between the ancestors of strains from Nigeria (see Fig. 5). Branches leading to most lineages outside of Nigeria appear to have diverged more recently, suggesting that the ancestral population of all modern Lassa virus lineages was in Nigeria. Hence, Lassa virus most likely emerged in present-day Nigeria, at least 1000 years ago (Ehichioya et al. 2011; Andersen et al. 2015; Manning et al. 2015; Oloniniyi et al. 2018). The virus subsequently spread within the country, giving rise to at least three major lineages (I, II, and III). Among them, lineage I has the earliest branching time while lineages II and III have shared a common ancestor at least ~600 years ago, indicating that lineage I likely emerged first, with lineages II and III being more recent (Ehichioya et al. 2011; Oloniniyi et al. 2018).

Estimates for the time of the major branching event between Nigerian and non-Nigerian lineages indicate a shared most recent common ancestor ~400 years ago, an estimate with low confidence since that divergence event is between lineage III and a single non-Nigerian sequence. Thus, 400 years ago at the latest, the virus would have come out of Nigeria and spread west to the countries of the Mano River Union (Guinea, Liberia, and Sierra Leone), the Ivory Coast, Mali, and possibly other regions in West Africa (Bowen et al. 2000; Andersen et al. 2015; Manning et al. 2015; Wiley et al. 2019). This event led to the emergence of at least one lineage out of Nigeria (IV) and likely more (V and VI).

Lineage IV includes sequences from the Mano River Union (MRU) region that share their most recent common ancestor at least ~300 years ago. Given the small number of sequences available from Guinea and Liberia, it is difficult to estimate where the MRCA of lineage IV actually circulated and describe the movement of the virus as it entered the MRU region (Wiley et al. 2019). Within lineage IV, Liberian genomes have the earliest branching time and are intermixed with genomes from Guinea (Fig. 5). The lack of clear separation between sequences from Liberia and Guinea suggest that the virus has been circulating between the two regions over the last 300 years. All currently available genomes from Sierra Leone cluster into a single clade that exclusively includes sequences from this country, with an estimated tMRCA of ~150 years, which is younger than that of clades including Guinean and Liberian sequences. Hence, based on the currently circulating genetic diversity, Lassa virus would have entered Sierra Leone more recently, around 150 years ago at the latest (Bowen et al. 2000; Andersen et al. 2015; Manning et al. 2015; Wiley et al. 2019). To date, the factors that supported the dispersal of the virus from Nigeria toward West Africa have not been identified and it is not clear whether the virus circulated and still circulates in other countries along that hypothetical path.

4.2 Regional Dynamics

The main Lassa virus lineages are associated with distinct geographic areas. An increasing number of studies indicate that additional local geographic structure is present, hinting that environmental factors may impact the circulation of the virus. Several regions that border Lassa endemic areas have reported isolated Lassa fever cases, and it is suspected that Lassa virus is circulating more widely, but little or no sequence data is available. Hence, increased sampling is needed to better capture both local and regional dynamics.

4.2.1 Nigeria

The number of Lassa virus sequences available from Nigeria has increased considerably over the last few years (Fig. 4b). These sequences fall into at least three major lineages. Lineage I is defined by the Pinneo strain, isolated in Borno State in 1969 (Buckley and Casals 1970), with no other identified sequence for this lineage, though Borno and the surrounding northeastern states are underrepresented among subsequently available Lassa virus sequences. Lineage II has been extensively documented and includes viruses collected from patients in Edo, Ondo, and Ebonyi States in southern Nigeria, while viruses collected in northern and central parts of the country are grouped into lineage III. A recent phylogeographic analysis suggests that lineage II originated in what is currently Ebonyi state and then spread to the West, while the ancestor that gave rise to lineage III appears to have spread in all directions from the North of Plateau State (Ehichioya et al. 2019).

These patterns suggest a role of geographic boundaries in the divergence of the Lassa virus. The separation between lineages II and III roughly aligns with the Benue River, while sublineages within lineage II correspond to opposite sides of the Niger River (Siddle et al. 2018; Ehichioya et al. 2019). Since most human infections are thought to be acquired from infected *Mastomys natalensis* rodents, the spatial spread of Lassa virus is primarily constrained by that of its rodent reservoir; the spatial specificity observed for Lassa virus lineages II and III is consistent with the limited range of *M. natalensis* (Mariën et al. 2018), with environmental barriers (e.g., waterways) further limiting dispersal.

4.2.2 Sierra Leone, Liberia and Guinea

Current knowledge of Lassa virus genetic diversity in Sierra Leone is driven by extensive sampling over the last decade and includes virus captured from both rodent samples and human cases. While the first published reports of Lassa fever in Sierra Leone date back to the early 1970s, the virus has likely been present in the country for much longer (Andersen et al. 2015; Wiley et al. 2019). As with Nigeria (Ehichioya et al. 2011), sequences in Sierra Leone lineage IV show geographic clustering. Phylogenetic analysis of a set of rodent sequences collected in 2009 in the eastern part of the country found significant spatial clustering at the regional level (Leski et al. 2015). Though they require confirmation with systematic sampling at more locations and time points, these estimates of temporal and spatial dynamics can inform rodent elimination strategies, identify regions at risk for new emergence of Lassa and help explain the life cycle of Lassa virus in its natural host.

Although regions of Guinea and Liberia are considered endemic for Lassa, the virus has been understudied in these countries compared to Sierra Leone and Nigeria, with 13 complete or nearly complete genomic sequences are available from Guinea and 25 from Liberia. In Guinea, serosurveys in humans show evidence of exposure in the southeastern forest regions close to the borders with Sierra Leone and Liberia, with seroprevalence similar to that observed in endemic areas of Sierra Leone (Lukashevich et al. 1993; Bausch et al. 2001; Kernéis et al. 2009). A twelve-year survey of Lassa virus from rodents in the municipality of Faranah in Upper Guinea provided a look at virus circulation on a local scale. The phylogenetic reconstruction using a set of 132 partial nucleoprotein genes shows considerable spatial specificity, with monophyletic clusters tending to map to the same village. However, polyphyletic clusters were also seen in some villages, and a few monophyletic clusters included sequences from different villages. These patterns should be confirmed by increasing both the extent and the homogeneity of the sampling, but suggest that on a local scale, the virus may circulate within and between villages (Fichet-Calvet et al. 2016). A similar approach that includes clinical and rodent samples from the same villages is needed to capture rodent-to-human transmission and further refine models of the zoonotic dynamics of Lassa fever (e.g., Viana et al. 2014; Lo Iacono et al. 2016; Plowright et al. 2017; Ambrose et al. 2019; Becker et al. 2019).

In Liberia, Lassa virus epidemiology has changed in recent years, as cases of Lassa fever have been reported South of the historical Lassa Belt that includes Lofa, Nimba, Bong, and Grand Gedeh Counties (Olugasa et al. 2015; Woyessa et al. 2019; Wiley et al. 2019), though these have often been isolated diagnoses with no documented secondary transmission. In some cases, phylogenetic analysis could be combined with epidemiological data to show that patients diagnosed in Montserrado had most likely been infected in Nimba County before seeking care in the South (Wiley et al. 2019). In September 2019, Liberia declared a new outbreak of Lassa fever with cases reported in Grand Bassa, Bong, Grand Kru, and Nimba Counties (2019b). Viral genomes from this outbreak are needed to understand whether cases in Grand Bassa and Grand Kru Counties are being imported or if Lassa virus is settling in the southern part of Liberia.

4.2.3 Elsewhere in West Africa

Lassa virus has been detected in a number of countries bordering on known Lassa endemic regions. Models incorporating the range of *M. natalensis* and environmental covariates suggest many of these regions have a suitable ecological niche for zoonotic transmission of Lassa virus (Fichet-Calvet and Rogers 2009; Mylne et al. 2015), although cases of human disease are rarely reported.

The few genomes sequenced from Mali and Côte d'Ivoire on the one hand, and from Togo on the other hand, suggest the existence of additional lineages, implying that Lassa virus is evolving in isolated reservoir populations in these regions. Since 2012, five complete or nearly complete genome sequences have been obtained from Mali and thirteen complete coding sequences from Côte d'Ivoire, from both rodent and human sources (Günther et al. 2000; Atkin et al. 2009; Safronetz et al. 2010; Manning et al. 2015). Most of these viral sequences are genetically distinct from those of lineage IV, leading to the proposal of a lineage V (Manning et al. 2015; Jan Baumann et al. 2019) that may have been circulating without detection for some time. According to tMRCA estimates, the common ancestor to lineage V was circulating at least 100 years ago, an estimate that might be pushed further back in time as more diversity from that lineage is sampled (Manning et al. 2015; Jan Baumann et al. 2019). The first two confirmed cases of Lassa fever in Togo were reported in early 2016 in the northern part of the country (Patassi et al. 2017). Subsequent sequencing of a cluster of cases with an epidemiological link to Togo led to the proposal of another lineage, lineage VI, more closely related to sequences in Nigeria than those in the Mano River Union (Whitmer et al. 2018). The position of this sequence in the phylogenetic tree, however, depended on the gene used for alignment: the L gene clustered with lineage II sequences (posterior probability ~1.0) while the GP and NP genes clustered with lineages I and VI, respectively (posterior probability ~0.7). Again, more sequences are needed to better understand Lassa virus circulation in Togo.

Other regions have evidence or suspicions of Lassa circulation, however with no usable sequence information. For viral sequences from travelers who visited multiple countries at risk for Lassa fever, the phylogenetic placement is difficult to interpret because the time and location of infection cannot be determined with certainty (Günther et al. 2000; Atkin et al. 2009), exemplified by sequences taken from Ghanaian soldiers on mission in Liberia (Bonney et al. 2016). Ghana and Benin have reported laboratory-confirmed Lassa fever cases, but again, without sequence data (Dzotsi et al. 2012, 2016, 2017). Other nearby countries are considered "at risk" for Lassa virus as they are home to *M. natalensis* populations and border endemic areas (e.g., Senegal, Guinea-Bissau, Niger, Cameroon) (Mylne et al. 2015). Active surveillance in rodents and potentially asymptomatic humans in these areas is needed, especially close to borders with endemic countries. With centuries of evolution and a widely distributed rodent reservoir, it seems likely that the extent and diversity of Lassa virus remains under-characterized. Increased intensity and breadth of surveillance are needed to more fully characterize Lassa virus circulation and zoonotic risk in West Africa.

Since 2014, the amount of available sequences has allowed notable improvement in our understanding of Lassa virus genetics, transmission, and historical spread. Future work should further leverage massively parallel sequencing to explore the dynamics of the virus genome within individual hosts over the course of an infection, better characterize environmental factors in virus transmission and evolution, and relate genetic diversity to clinical outcomes. Understanding the virus at the population level is, however, biased by the variability in clinical, public health, and scientific resources available between countries. Systematic and more widespread surveillance is needed in both human and rodent populations to reduce these biases and fill in gaps in our knowledge.

5 Clinical Implications

Beyond phylodynamics, genomic analysis can identify mutations that influence viral phenotype, epidemiology, and pathophysiology (Diehl et al. 2016; Geoghegan and Holmes 2018). While there is suggestive evidence for phenotypic variation among Lassa lineages, the strongest data come from laboratory studies which may not translate accurately to human disease, since humans are usually a dead-end host and the main selective pressure on Lassa virus is the rodent reservoir. Human studies further face extensive confounds from human genetic backgrounds, variability in medical care, and socioeconomic context. Genetic diversity can, however, translate into antigenic diversity that must be considered in the development of vaccines and passive antibody therapies.

5.1 Genetics and Phenotype

Between 2008 and 2013, the observed case fatality rates for Lassa fever were markedly higher at Kenema Government Hospital in Sierra Leone compared to Irrua Specialist Teaching Hospital in Nigeria. Despite similar patient presentation times and available treatments, viral loads were also observed to be higher in Kenema than in Irrua (Andersen et al. 2015), raising the possibility that the respective Lassa lineages differ in virulence.

In vivo laboratory studies have compared pathogenicity of Lassa virus strains in guinea pigs and cynomolgus macaques (Jahrling et al. 1985b; Safronetz et al. 2013, 2015; Welch et al. 2018). In both tested animal species, more severe infections were reported with Josiah (Wulff and Johnson 1979) and Soromba-R (Safronetz et al. 2010) strains than with the Liberian strain Z-132 (Jahrling et al. 1985b). Since these studies used cultured strains with varying passage histories, it is not clear whether they reflect inter-lineage differences, particularly since substantial differences in virulence have also been reported within a single lineage (Jahrling et al. 1985b). Other studies have examined the pathogenicity of single Lassa strains in isolation (e.g., Peters et al. 1987), but variability in model systems and protocols complicates comparison across studies. More systematic and detailed investigations in both human and animal hosts are needed to understand how the evolution of the Lassa virus genome may impact its phenotype and whether it influences critical parameters of the disease, such as transmission or severity in humans.

5.2 Genetics and Vaccines

The WHO's goals for Lassa vaccine candidates highlights the importance of protection across viral lineages and efficacy following single immunization, given logistical and cold chain constraints in West Africa (2018). These requirements argue for a live-virus vaccine to induce a strong immune response, although it remains unclear whether pan-Lassa protection is achievable given the genetic diversity of the virus. In natural human infection, the impact of viral genomic diversity on cross-lineage protection can be assessed by studying viral escape mutations. Andersen et al. (2015) found that human clinical samples include non-synonymous mutations that are enriched in predicted immune epitopes compared to synonymous mutations (Andersen et al. 2015). This was particularly notable for epitopes in GPC, the only exposed surface protein on a Lassa virion, and some of these variants can disrupt binding of the potent pan-Lassa neutralizing antibody 12.1F (Robinson et al. 2016), highlighting the potential for immune escape and decreased vaccine efficacy. Since human-to-human transmission is rare, however, immune escape during human infection is unlikely to give rise to circulating vaccine-resistant strains.

Cross-lineage protection in experimental vaccination studies has not been thoroughly explored thus far. Recently, however, Cross et al. (2020) described a quadrivalent recombinant-VSV vaccine that expressed the GP genes of the Ebola, Sudan, and Marburg filoviruses and lineage IV Josiah Lassa virus. The vaccine-induced neutralizing antibodies in all studied macaques following two immunizations and protected them from both homologous challenge with corresponding filoviruses and heterologous challenge with a lineage II Lassa virus isolate from Nigeria. Whether the induced immune responses are protective against other lineages remains to be seen. While viral vaccines typically rely on antibody responses, a number of recent vaccine candidates, including those for highly variable pathogens like HIV, protect through CD8+ T cell responses (Hansen et al. 2016; Lukashevich et al. 2019). This approach is appealing for Lassa virus, as clearance of primary Lassa virus and LCMV infection is associated with T cell-mediated immunity (Prescott et al. 2017). However, even less is known about the cross-protectiveness of vaccine-induced T cell responses compared to antibody-mediated immune correlates. One vaccine candidate, a Mopeia/Lassa reassortment live-virus vaccine called ML29, appears to induce cell-mediated immunity in non-human primates that protects from Lassa virus strains derived from both Sierra Leone and Nigeria (Lukashevich et al. 2008, 2019). Most other candidates with robust cellular and/or humoral immune responses have not been assessed for cross-protection against diverse Lassa strains (Lukashevich et al. 2008).

Serological studies following natural infection, however, can likely give an indication of how readily pan-Lassa protection may be achievable, provided that effective vaccines closely mimicking natural infection can be developed. Early reports by Jahrling et al. (1985b) did demonstrate that polyclonal immune sera from convalescent Sierra Leonean and Liberian patients can cross-neutralize Liberian and Sierra Leonean virus isolates, respectively, although both had lower potency against Nigerian strains. Convalescent sera also provided passive protection in a guinea pig model, although cross-lineage protection was not systematically assessed (Jahrling et al. 1985a). More recent studies specifically considering cross-lineage neutralization have found several antibodies that are broadly neutralizing against diverse Lassa virus lineages (Robinson et al. 2016). Interestingly, when Hastie et al. (2019) investigated structural features of broadly neutralizing monoclonal antibodies isolated from Lassa fever survivors, they found that a relatively small number of mutations were required to develop pan-Lassa neutralization from germline antibody precursors.

Taken together, these results suggest that typical humoral immune responses to infection may be able to overcome the antigenic diversity of Lassa virus, but future studies will have to show if the potency of convalescent sera is high enough to protect from diverse lineages, and if vaccine-induced immune responses can induce similarly broad and potent immune correlates as natural infection. While vaccines are still under development, recombinant pan-Lassa broadly neutralizing antibodies can also serve in passive antibody therapy, akin to recently approved antibody-based Ebola therapies (Cross et al. 2019). In the case of Lassa virus, a mixture of pan-Lassa neutralizing antibodies targeting different epitopes may help address both genetic diversity and the possibility of escape mutations.

6 Conclusions

The past decade has seen shifts in the numbers and geography of reported cases of Lassa fever. These shifts may be multifactorial, with contributions ranging from changing reservoir ecology to viral adaptation to evolving human awareness. The Lassa virus genome can reflect signatures of each of these factors, and when combined with traditional ecology and virology, will both expand our understanding of this virus and pave the way for future applications.

References

Albariño CG, Bergeron E, Erickson BR et al (2009) Efficient reverse genetics generation of infectious junin viruses differing in glycoprotein processing. J Virol 83:5606–5614

Albariño CG, Bird BH, Chakrabarti AK et al (2011) Efficient rescue of recombinant Lassa virus reveals the influence of S segment noncoding regions on virus replication and virulence. J Virol 85:4020–4024

Ambrose MR, Kucharski AJ, Formenty P et al (2019) Quantifying transmission of emerging zoonoses: using mathematical models to maximize the value of surveillance data. bioRxiv 677021

Andersen KG, Shapiro BJ, Matranga CB et al (2015) Clinical sequencing uncovers origins and evolution of Lassa virus. Cell 162:738–750

Archer AM, Rico-Hesse R (2002) High genetic divergence and recombination in Arenaviruses from the Americas. Virology 304:274–281

Armstrong C, Lillie RD (1934) US Publ. Hlth Rep 49:1019

Asogun DA, Adomeh DI, Ehimuan J et al (2012) Molecular diagnostics for Lassa fever at Irrua specialist teaching hospital, Nigeria: lessons learnt from two years of laboratory operation. PLoS Negl Trop Dis 6:e1839

Atkin S, Anaraki S, Gothard P et al (2009) The first case of Lassa fever imported from Mali to the United Kingdom, February 2009. Euro Surveill 14

Auperin DD, McCormick JB (1989) Nucleotide sequence of the Lassa virus (Josiah strain) S genome RNA and amino acid sequence comparison of the N and GPC proteins to other arenaviruses. Virology 168:421–425

Auperin DD, Romanowski V, Galinski M, Bishop DH (1984) Sequencing studies of pichinde arenavirus S RNA indicate a novel coding strategy, an ambisense viral S RNA. J Virol 52:897–904

Ayling M, Clark MD, Leggett RM (2019) New approaches for metagenome assembly with short reads. Brief Bioinform. https://doi.org/10.1093/bib/bbz020

Baumann J, Knüpfer M, Ouedraogo J et al (2019) Lassa and Crimean-Congo hemorrhagic Fever Viruses, Mali. Emerg Infect Dis J 25:999

Bausch DG, Rollin PE, Demby AH et al (2000) Diagnosis and clinical virology of Lassa fever as evaluated by enzyme-linked immunosorbent assay, indirect fluorescent-antibody test, and virus isolation. J Clin Microbiol 38:2670–2677

Bausch DG, Demby AH, Coulibaly M et al (2001) Lassa fever in Guinea: I. Epidemiology of human disease and clinical observations. Vector Borne Zoonotic Dis 1:269–281

Becker DJ, Washburne AD, Faust CL et al (2019) Dynamic and integrative approaches to understanding pathogen spillover. Philos Trans R Soc Lond B Biol Sci 374:20190014

Beitzel B, Hulseberg CE, Palacios G (2019) Reverse genetics systems as tools to overcome the genetic diversity of Lassa virus. Curr Opin Virol 37:91–96

Bergeron É, Chakrabarti AK, Bird BH et al (2012) Reverse genetics recovery of Lujo virus and role of virus RNA secondary structures in efficient virus growth. J Virol 86:10759–10765
Boisen ML, Hartnett JN, Shaffer JG et al (2018) Field validation of recombinant antigen immunoassays for diagnosis of Lassa fever. Sci Rep 8:5939
Bonney JHK, Nyarko EO, Ohene S-A et al (2016) Molecular confirmation of Lassa fever imported into Ghana. S Afr J Lab Clin Med 5:288
Both GW, Furuichi Y, Muthukrishnan S, Shatkin AJ (1975) Ribosome binding to reovirus mRNA in protein synthesis requires 5' terminal 7-methylguanosine. Cell 6:185–195
Bowen MD, Rollin PE, Ksiazek TG et al (2000) Genetic diversity among Lassa virus strains. J Virol 74:6992–7004
Branco LM, Grove JN, Geske FJ et al (2010) Lassa virus-like particles displaying all major immunological determinants as a vaccine candidate for Lassa hemorrhagic fever. Virol J 7:279
Branco LM, Grove JN, Boisen ML et al (2011) Emerging trends in Lassa fever: redefining the role of immunoglobulin M and inflammation in diagnosing acute infection. Virol J 8:478
Briese T, Paweska JT, McMullan LK et al (2009) Genetic detection and characterization of Lujo virus, a new hemorrhagic fever-associated arenavirus from southern Africa. PLoS Pathog 5: e1000455
Briese T, Kapoor A, Mishra N et al (2015) Virome capture sequencing enables sensitive viral diagnosis and comprehensive Virome analysis. MBio 6:e01491–15
Brisse ME, Ly H (2019) Hemorrhagic fever-causing Arenaviruses: lethal pathogens and potent immune suppressors. Front Immunol 10:372
Brunotte L, Lelke M, Hass M et al (2011) Domain structure of Lassa virus L protein. J Virol 85:324–333
Bruns M, Lehmann-Grube F (1983) Lymphocytic choriomeningitis virus. V. Proposed structural arrangement of proteins in the virion. J Gen Virol 64(Pt 10):2157–2167
Buckley SM, Casals J (1970) Lassa fever, a new virus disease of man from West Africa. 3. Isolation and characterization of the virus. Am J Trop Med Hyg 19:680–691
Burri DJ, da Palma JR, Kunz S, Pasquato A (2012) Envelope glycoprotein of arenaviruses. Viruses 4:2162–2181
Caì Y, Iwasaki M, Beitzel BF et al (2018) Recombinant Lassa virus expressing green fluorescent protein as a tool for high-throughput drug screens and neutralizing antibody assays. Viruses 10. https://doi.org/10.3390/v10110655
Cao W, Henry MD, Borrow P et al (1998) Identification of alpha-dystroglycan as a receptor for lymphocytic choriomeningitis virus and Lassa fever virus. Science 282:2079–2081
Capul AA, de la Torre JC, Buchmeier MJ (2011) Conserved residues in Lassa fever virus Z protein modulate viral infectivity at the level of the ribonucleoprotein. J Virol 85:3172–3178
Carey DE, Kemp GE, White HA et al (1972) Lassa fever. Epidemiological aspects of the 1970 epidemic, Jos, Nigeria. Trans R Soc Trop Med Hyg 66:402–408
Carnec X, Baize S, Reynard S et al (2011) Lassa virus nucleoprotein mutants generated by reverse genetics induce a robust type I interferon response in human dendritic cells and macrophages. J Virol 85:12093–12097
Casabona JC, Levingston Macleod JM, Loureiro ME et al (2009) The RING domain and the L79 residue of Z protein are involved in both the rescue of nucleocapsids and the incorporation of glycoproteins into infectious chimeric arenavirus-like particles. J Virol 83:7029–7039
Charrel RN, de Lamballerie X (2003) Arenaviruses other than Lassa virus. Antiviral Res 57:89–100
Charrel RN, de Lamballerie X, Fulhorst CF (2001) The whitewater Arroyo virus: natural evidence for genetic recombination among Tacaribe serocomplex viruses (family Arenaviridae). Virology 283:161–166
Charrel RN, Feldmann H, Fulhorst CF et al (2002) Phylogeny of new world arenaviruses based on the complete coding sequences of the small genomic segment identified an evolutionary lineage produced by intrasegmental recombination. Biochem Biophys Res Commun 296:1118–1124

Chen JS, Ma E, Harrington LB et al (2018) CRISPR-Cas12a target binding unleashes indiscriminate single-stranded DNase activity. Science 360:436–439
Cohen-Dvashi H, Israeli H, Shani O et al (2016) Role of LAMP1 binding and pH sensing by the spike complex of Lassa virus. J Virol 90:10329–10338
Cosset F-L, Marianneau P, Verney G et al (2009) Characterization of Lassa virus cell entry and neutralization with Lassa virus pseudoparticles. J Virol 83:3228–3237
Cross RW, Hastie KM, Mire CE et al (2019) Antibody therapy for Lassa fever. Curr Opin Virol 37:97–104
Cross RW, Xu R, Matassov D et al (2020) Quadrivalent VesiculoVax vaccine protects nonhuman primates from viral-induced hemorrhagic fever and death. J Clin Invest 130:539–551
Das S, Rundell MS, Mirza AH et al (2015) A multiplex PCR/LDR assay for the simultaneous identification of category A infectious pathogens: agents of viral hemorrhagic fever and Variola virus. PLoS ONE 10:e0138484
Demby AH, Chamberlain J, Brown DW, Clegg CS (1994) Early diagnosis of Lassa fever by reverse transcription-PCR. J Clin Microbiol 32:2898–2903
Dhanwani R, Zhou Y, Huang Q et al (2015) A novel live Pichinde Virus-based vaccine vector induces enhanced humoral and cellular immunity after a booster dose. J Virol 90:2551–2560
Diehl WE, Lin AE, Grubaugh ND et al (2016) Ebola virus glycoprotein with increased infectivity dominated the 2013–2016 epidemic. Cell 167:1088–1098.e6
Djavani M, Lukashevich IS, Sanchez A et al (1997) Completion of the Lassa fever virus sequence and identification of a RING finger open reading frame at the L RNA 5' End. Virology 235:414–418
Drummond AJ, Suchard MA, Xie D, Rambaut A (2012) Bayesian phylogenetics with BEAUti and the BEAST 1.7. Mol Biol Evol 29:1969–1973
Duarte HA, Panpradist N, Beck IA et al (2017) Current status of point-of-care testing for human immunodeficiency virus drug resistance. J Infect Dis 216:S824–S828
Dzotsi EK, Ohene S-A, Asiedu-Bekoe F et al (2012) The first cases of Lassa fever in Ghana. Ghana Med J 46:166–170
Eberhardt KA, Mischlinger J, Jordan S et al (2019) Ribavirin for the treatment of Lassa fever: a systematic review and meta-analysis. Int J Infect Dis 87:15–20
Ehichioya DU, Hass M, Becker-Ziaja B et al (2011) Current molecular epidemiology of Lassa virus in Nigeria. J Clin Microbiol 49:1157–1161
Ehichioya DU, Dellicour S, Pahlmann M et al (2019) Phylogeography of Lassa virus in Nigeria. J Virol. https://doi.org/10.1128/JVI.00929-19
Eichler R, Lenz O, Strecker T, Garten W (2003) Signal peptide of Lassa virus glycoprotein GP-C exhibits an unusual length. FEBS Lett 538:203–206
Emonet SF, Garidou L, McGavern DB, de la Torre JC (2009) Generation of recombinant lymphocytic choriomeningitis viruses with trisegmented genomes stably expressing two additional genes of interest. Proc Natl Acad Sci U S A 106:3473–3478
Emonet SF, Seregin AV, Yun NE et al (2011) Rescue from cloned cDNAs and in vivo characterization of recombinant pathogenic Romero and live-attenuated Candid #1 strains of Junin virus, the causative agent of Argentine hemorrhagic fever disease. J Virol 85:1473–1483
Eschli B, Quirin K, Wepf A et al (2006) Identification of an N-terminal trimeric coiled-coil core within arenavirus glycoprotein 2 permits assignment to class I viral fusion proteins. J Virol 80:5897–5907
Faria NR, Kraemer MUG, Hill SC et al (2018) Genomic and epidemiological monitoring of yellow fever virus transmission potential. Science 361:894–899
Fedeli C, Torriani G, Galan-Navarro C et al (2018) Axl can serve as entry factor for Lassa virus depending on the functional Glycosylation of Dystroglycan. J Virol 92. https://doi.org/10.1128/JVI.01613-17
Ferron F, Weber F, de la Torre JC, Reguera J (2017) Transcription and replication mechanisms of Bunyaviridae and Arenaviridae L proteins. Virus Res 234:118–134
Fichet-Calvet E, Rogers DJ (2009) Risk maps of Lassa fever in West Africa. PLoS Negl Trop Dis 3:e388

Fichet-Calvet E, Ölschläger S, Strecker T et al (2016) Spatial and temporal evolution of Lassa virus in the natural host population in Upper Guinea. Sci Rep 6:21977

Flatz L, Bergthaler A, de la Torre JC, Pinschewer DD (2006) Recovery of an arenavirus entirely from RNA polymerase I/II-driven cDNA. Proc Natl Acad Sci U S A 103:4663–4668

Flint J, Racaniello VR, Rall GF et al (eds) (2015) Chapter 5: Mechanisms of pathogenesis. In: Principles of virology, Fourth Edition, Volume II. American Society of Microbiology, pp 134–173

Forni D, Pontremoli C, Pozzoli U et al (2018) Ancient evolution of Mammarenaviruses: adaptation via changes in the L protein and no evidence for host-virus codivergence. Genome Biol Evol 10:863–874

Foscaldi S, D'Antuono A, Noval MG et al (2017) Regulation of tacaribe Mammarenavirus translation: positive 5' and negative 3' elements and role of key cellular factors. J Virol 91. https://doi.org/10.1128/JVI.00084-17

Fukuma A, Kurosaki Y, Morikawa Y et al (2011) Rapid detection of Lassa virus by reverse transcription-loop-mediated isothermal amplification. Microbiol Immunol 55:44–50

Fuller-Pace FV, Southern PJ (1988) Temporal analysis of transcription and replication during acute infection with lymphocytic choriomeningitis virus. Virology 162:260–263

Furuichi Y, LaFiandra A, Shatkin AJ (1977) 5'-terminal structure and mRNA stability. Nature 266:235–239

Garcin D, Kolakofsky D (1992) Tacaribe arenavirus RNA synthesis in vitro is primer dependent and suggests an unusual model for the initiation of genome replication. J Virol 66:1370–1376

Gardy JL, Loman NJ (2018) Towards a genomics-informed, real-time, global pathogen surveillance system. Nat Rev Genet 19:9–20

Gaudin R, Kirchhausen T (2015) Superinfection exclusion is absent during acute Junin virus infection of Vero and A549 cells. Sci Rep 5:15990

Geisbert TW, Jones S, Fritz EA et al (2005) Development of a new vaccine for the prevention of Lassa fever. PLoS Med 2:e183

Geoghegan JL, Holmes EC (2018) The phylogenomics of evolving virus virulence. Nat Rev Genet 19:756–769

Gire SK, Goba A, Andersen KG et al (2014) Genomic surveillance elucidates Ebola virus origin and transmission during the 2014 outbreak. Science 345:1369–1372

Godfrey C, Clement E, Mein R et al (2007) Refining genotype phenotype correlations in muscular dystrophies with defective glycosylation of dystroglycan. Brain 130:2725–2735

Goncalves A-R, Moraz M-L, Pasquato A et al (2013) Role of DC-SIGN in Lassa virus entry into human dendritic cells. J Virol 87:11504–11515

Gonzalez JP, McCormick JB, Kiley MP (1988) Genetic variation among Lassa and Lassa-related arenaviruses analysed by T1-oligonucleotide mapping. Ann Inst Pasteur Virol 139:405–420

Gootenberg JS, Abudayyeh OO, Lee JW et al (2017) Nucleic acid detection with CRISPR-Cas13a/C2c2. Science 356:438–442

Grubaugh ND, Ladner JT, Lemey P et al (2019) Tracking virus outbreaks in the twenty-first century. Nat Microbiol 4:10–19

Gryseels S, Baird SJE, Borremans B et al (2017) When viruses don't go viral: the importance of host phylogeographic structure in the spatial spread of Arenaviruses. PLoS Pathog 13: e1006073

Günther S, Emmerich P, Laue T et al (2000) Imported Lassa fever in Germany: molecular characterization of a new Lassa virus strain. Emerg Infect Dis 6:466–476

Hallam SJ, Koma T, Maruyama J, Paessler S (2018) Review of Mammarenavirus biology and replication. Front Microbiol 9:1751

Hansen SG, Wu HL, Burwitz BJ, Hughes CM (2016) Broadly targeted CD8+ T cell responses restricted by major histocompatibility complex E

Hara Y, Kanagawa M, Kunz S et al (2011) Like-acetylglucosaminyltransferase (LARGE)-dependent modification of dystroglycan at Thr-317/319 is required for laminin binding and Arenavirus infection. Proc Natl Acad Sci U S A 108:17426–17431

Hass M, Gölnitz U, Müller S et al (2004) Replicon system for Lassa virus. J Virol 78:13793–13803
Hass M, Westerkofsky M, Müller S et al (2006) Mutational analysis of the Lassa virus promoter. J Virol 80:12414–12419
Hastie KM, Saphire EO (2018) Lassa virus glycoprotein: stopping a moving target. Curr Opin Virol 31:52–58
Hastie KM, Cross RW, Harkins SS et al (2019) Convergent structures illuminate features for germline antibody binding and Pan-Lassa virus neutralization. Cell 178:1004–1015.e14
Hepojoki J, Hepojoki S, Smura T et al (2018) Characterization of Haartman Institute snake virus-1 (HISV-1) and HISV-like viruses—The representatives of genus Hartmanivirus, family Arenaviridae. PLoS Pathog 14:e1007415
Houldcroft CJ, Beale MA, Breuer J (2017) Clinical and biological insights from viral genome sequencing. Nat Rev Microbiol 15:183–192
Huang Q, Shao J, Liang Y, Ly H (2018) Assays to demonstrate the roles of Arenaviral Nucleoproteins (NPs) in viral RNA synthesis and in suppressing type I interferon. Methods Mol Biol 1604:189–200
Hui DS, Azhar EI, Madani TA et al (2020) The continuing 2019-nCoV epidemic threat of novel coronaviruses to global health—the latest 2019 novel Corona virus outbreak in Wuhan, China. Int J Infect Dis 91:264–266
Iwasaki M, Minder P, Caì Y et al (2018) Interactome analysis of the lymphocytic choriomeningitis virus nucleoprotein in infected cells reveals ATPase Na+/K+ transporting subunit Alpha 1 and prohibitin as host-cell factors involved in the life cycle of mammarenaviruses. PLoS Pathog 14: e1006892
Jahrling PB, Frame JD, Rhoderick JB, Monson MH (1985a) Endemic Lassa fever in Liberia. IV. Selection of optimally effective plasma for treatment by passive immunization. Trans R Soc Trop Med Hyg 79:380–384
Jahrling PB, Frame JD, Smith SB, Monson MH (1985b) Endemic Lassa fever in Liberia. III. Characterization of Lassa virus isolates. Trans R Soc Trop Med Hyg 79:374–379
Jenkins GM, Rambaut A, Pybus OG, Holmes EC (2002) Rates of molecular evolution in RNA viruses: a quantitative phylogenetic analysis. J Mol Evol 54:156–165
Johnson DM, Jokinen JD, Lukashevich IS (2019) Attenuated replication of Lassa virus vaccine candidate ML29 in STAT-1-/- Mice. Pathogens 8. https://doi.org/10.3390/pathogens8010009
Kafetzopoulou LE, Pullan ST, Lemey P et al (2019) Metagenomic sequencing at the epicenter of the Nigeria 2018 Lassa fever outbreak. Science 363:74–77
Kainulainen MH, Spengler JR, Welch SR et al (2018) Use of a scalable replicon-particle vaccine to protect against lethal Lassa Virus infection in the Guinea pig model. J Infect Dis 217:1957–1966
Kane M, Golovkina T (2010) Common threads in persistent viral infections. J Virol 84:4116–4123
Kernéis S, Koivogui L, Magassouba N et al (2009) Prevalence and risk factors of Lassa seropositivity in inhabitants of the forest region of Guinea: a cross-sectional study. PLoS Negl Trop Dis 3:e548
King AMQ, Adams MJ, Carstens EB, Lefkowitz EJ (eds) (2012) Family—Arenaviridae. In: Virus taxonomy. Elsevier, San Diego, pp 715–723
King BR, Hershkowitz D, Eisenhauer PL et al (2017) A map of the Arenavirus nucleoprotein-host protein interactome reveals that Junín Virus selectively impairs the antiviral activity of double-stranded RNA-Activated Protein Kinase (PKR). J Virol 91. https://doi.org/10.1128/JVI.00763-17
King BR, Samacoits A, Eisenhauer PL et al (2018) Visualization of Arenavirus RNA species in individual cells by single-molecule fluorescence in situ hybridization suggests a model of cyclical infection and clearance during persistence. J Virol 92. https://doi.org/10.1128/JVI.02241-17
Koehler JW, Douglas CE, Minogue TD (2018) A highly multiplexed broad pathogen detection assay for infectious disease diagnostics. PLoS Negl Trop Dis 12:e0006889

Kronmann KC, Nimo-Paintsil S, Guirguis F et al (2013) Two novel arenaviruses detected in pygmy mice, Ghana. Emerg Infect Dis 19:1832–1835
Kunz S, Rojek JM, Kanagawa M et al (2005a) Posttranslational modification of alpha-dystroglycan, the cellular receptor for Arenaviruses, by the glycosyltransferase LARGE is critical for virus binding. J Virol 79:14282–14296
Kunz S, Rojek JM, Perez M et al (2005b) Characterization of the interaction of Lassa fever virus with its cellular receptor alpha-dystroglycan. J Virol 79:5979–5987
Lan S, McLay Schelde L, Wang J et al (2009) Development of infectious clones for virulent and avirulent pichinde viruses: a model virus to study Arenavirus-induced hemorrhagic fevers. J Virol 83:6357–6362
Lee KJ, Novella IS, Teng MN et al (2000) NP and L proteins of lymphocytic choriomeningitis virus (LCMV) are sufficient for efficient transcription and replication of LCMV genomic RNA analogs. J Virol 74:3470–3477
Lee AM, Rojek JM, Spiropoulou CF et al (2008) Unique small molecule entry inhibitors of hemorrhagic fever arenaviruses. J Biol Chem 283:18734–18742
Lee AM, Cruite J, Welch MJ et al (2013) Pathogenesis of Lassa fever virus infection: I. Susceptibility of mice to recombinant Lassa Gp/LCMV chimeric virus. Virology 442:114–121
Lelke M, Brunotte L, Busch C, Günther S (2010) An N-terminal region of Lassa virus L protein plays a critical role in transcription but not replication of the virus genome. J Virol 84:1934–1944
Lenz O, ter Meulen J, Klenk HD et al (2001) The Lassa virus glycoprotein precursor GP-C is proteolytically processed by subtilase SKI-1/S1P. Proc Natl Acad Sci U S A 98:12701–12705
Leski TA, Stockelman MG, Moses LM et al (2015) Sequence variability and geographic distribution of Lassa virus, Sierra Leone. Emerg Infect Dis 21:609–618
Li K, Lin X-D, Wang W et al (2015) Isolation and characterization of a novel Arenavirus harbored by Rodents and Shrews in Zhejiang province, China. Virology 476:37–42
Li S, Sun Z, Pryce R et al (2016) Acidic pH-induced conformations and LAMP1 binding of the Lassa Virus glycoprotein spike. PLoS Pathog 12:e1005418
Li Q, Liu Q, Huang W et al (2017) An LASV GPC pseudotyped virus based reporter system enables evaluation of vaccines in mice under non-BSL-4 conditions. Vaccine 35:5172–5178
Lo Iacono G, Cunningham AA, Fichet-Calvet E et al (2016) A unified framework for the infection dynamics of Zoonotic spillover and spread. PLoS Negl Trop Dis 10:e0004957
López N, Jácamo R, Franze-Fernández MT (2001) Transcription and RNA replication of Tacaribe virus genome and antigenome analogs require N and L proteins: Z protein is an inhibitor of these processes. J Virol 75:12241–12251
Loureiro ME, Wilda M, Levingston Macleod JM et al (2011) Molecular determinants of arenavirus Z protein homo-oligomerization and L polymerase binding. J Virol 85:12304–12314
Loureiro ME, D'Antuono A, Levingston Macleod JM, López N (2012) Uncovering viral protein-protein interactions and their role in Arenavirus life cycle. Viruses 4:1651–1667
Loureiro ME, D'Antuono A, López N (2019) Virus-host interactions involved in Lassa virus entry and genome replication. Pathogens 8. https://doi.org/10.3390/pathogens8010017
Lukashevich IS (1992) Generation of reassortants between African arenaviruses. Virology 188:600–605
Lukashevich IS, Pushko P (2016) Vaccine platforms to control Lassa fever. Expert Rev Vaccines 15:1135–1150
Lukashevich IS, Stelmakh TA, Golubev VP et al (1984) Ribonucleic acids of Machupo and Lassa viruses. Arch Virol 79:189–203
Lukashevich IS, Clegg JC, Sidibe K (1993) Lassa virus activity in Guinea: distribution of human antiviral antibody defined using enzyme-linked immunosorbent assay with recombinant antigen. J Med Virol 40:210–217
Lukashevich IS, Djavani M, Shapiro K, Sanchez A, Ravkov E, Nichol ST, Salvato MS (1997) The Lassa fever virus L gene: nucleotide sequence, comparison, and precipitation of a predicted 250 kDa protein with monospecific antiserum. J Gen Virol 78(Pt 3):547–551

Lukashevich IS, Patterson J, Carrion R et al (2005) A live attenuated vaccine for Lassa fever made by reassortment of Lassa and Mopeia viruses. J Virol 79:13934–13942
Lukashevich IS, Carrion R Jr, Salvato MS et al (2008) Safety, immunogenicity, and efficacy of the ML29 reassortant vaccine for Lassa fever in small non-human primates. Vaccine 26:5246–5254
Lukashevich IS, Paessler S, de la Torre JC (2019) Lassa virus diversity and feasibility for universal prophylactic vaccine. F1000Res 8. https://doi.org/10.12688/f1000research.16989.1
Lunkenheimer K, Hufert FT, Schmitz H (1990) Detection of Lassa virus RNA in specimens from patients with Lassa fever by using the polymerase chain reaction. J Clin Microbiol 28:2689–2692
Ly H (2017) Differential immune responses to new world and old world mammalian Arenaviruses. Int J Mol Sci 18. https://doi.org/10.3390/ijms18051040
Madu IG, Files M, Gharaibeh DN et al (2018) A potent Lassa virus antiviral targets an Arenavirus virulence determinant. PLoS Pathog 14:e1007439
Maes P, Alkhovsky SV, Bào Y et al (2018) Taxonomy of the family Arenaviridae and the order Bunyavirales: update 2018. Arch Virol 163:2295–2310
Maes P, Adkins S, Alkhovsky SV et al (2019) Taxonomy of the order Bunyavirales: second update 2018. Arch Virol 164:927–941
Manning JT, Forrester N, Paessler S (2015) Lassa virus isolates from Mali and the Ivory Coast represent an emerging fifth lineage. Front Microbiol 6:1037
Mariën J, Borremans B, Gryseels S et al (2017) No measurable adverse effects of Lassa, Morogoro and Gairo arenaviruses on their rodent reservoir host in natural conditions. Parasit Vectors 10:210
Mariën J, Kourouma F, Magassouba N et al (2018) Movement patterns of small rodents in Lassa fever-endemic villages in Guinea. EcoHealth 15:348–359
Marq J-B, Kolakofsky D, Garcin D (2010) Unpaired 5' ppp-nucleotides, as found in Arenavirus double-stranded RNA panhandles, are not recognized by RIG-I. J Biol Chem 285:18208–18216
Marq J-B, Hausmann S, Veillard N et al (2011) Short double-stranded RNAs with an overhanging 5' ppp-nucleotide, as found in arenavirus genomes, act as RIG-I decoys. J Biol Chem 286:6108–6116
Martínez-Sobrido L, de la Torre JC (2017) Development of recombinant Arenavirus-based vaccines. Methods Mol Biol 1581:133–149
Matranga CB, Andersen KG, Winnicki S et al (2014) Enhanced methods for unbiased deep sequencing of Lassa and Ebola RNA viruses from clinical and biological samples. Genome Biol 15:519
Matranga CB, Gladden-Young A, Qu J et al (2016) Unbiased deep sequencing of RNA viruses from clinical samples. J Vis Exp. https://doi.org/10.3791/54117
Mazzola LT, Kelly-Cirino C (2019) Diagnostics for Lassa fever virus: a genetically diverse pathogen found in low-resource settings. BMJ Glob Health 4:e001116
McCormick JB (1987) Epidemiology and control of Lassa fever. Curr Top Microbiol Immunol 134:69–78
McCormick JB, Fisher-Hoch SP (2002) Lassa fever. Curr Top Microbiol Immunol 262:75–109
McCormick JB, King IJ, Webb PA et al (1986) Lassa fever. Effective therapy with ribavirin. N Engl J Med 314:20–26
McMullan LK, Flint M, Chakrabarti A et al (2019) Characterisation of infectious Ebola virus from the ongoing outbreak to guide response activities in the Democratic Republic of the Congo: a phylogenetic and in vitro analysis. Lancet Infect Dis. https://doi.org/10.1016/S1473-3099(19)30291-9
Metsky HC, Siddle KJ, Gladden-Young A et al (2019) Capturing sequence diversity in metagenomes with comprehensive and scalable probe design. Nat Biotechnol 37:160–168
Meyer BJ, Southern PJ (1993) Concurrent sequence analysis of 5' and 3' RNA termini by intramolecular circularization reveals 5' nontemplated bases and 3' terminal heterogeneity for lymphocytic choriomeningitis virus mRNAs. J Virol 67:2621–2627

Meyer BJ, de la Torre JC, Southern PJ (2002) Arenaviruses: genomic RNAs, transcription, and replication. Curr Top Microbiol Immunol 262:139–157

Mohr EL, McMullan LK, Lo MK et al (2015) Inhibitors of cellular kinases with broad-spectrum antiviral activity for hemorrhagic fever viruses. Antiviral Res 120:40–47

Monath TP, Mertens PE, Patton R et al (1973) A hospital epidemic of Lassa fever in Zorzor, Liberia, March-April 1972. Am J Trop Med Hyg 22:773–779

Myhrvold C, Freije CA, Gootenberg JS et al (2018) Field-deployable viral diagnostics using CRISPR-Cas13. Science 360:444–448

Mylne AQN, Pigott DM, Longbottom J et al (2015) Mapping the zoonotic niche of Lassa fever in Africa. Trans R Soc Trop Med Hyg 109:483–492

Naveca FG, Claro I, Giovanetti M et al (2019) Genomic, epidemiological and digital surveillance of Chikungunya virus in the Brazilian Amazon. PLoS Negl Trop Dis 13:e0007065

Neumann G, Watanabe T, Ito H et al (1999) Generation of influenza A viruses entirely from cloned cDNAs. Proc Natl Acad Sci U S A 96:9345–9350

Nikisins S, Rieger T, Patel P et al (2015) International external quality assessment study for molecular detection of Lassa virus. PLoS Negl Trop Dis 9:e0003793

Ogbu O, Ajuluchukwu E, Uneke CJ (2007) Lassa fever in West African sub-region: an overview. J Vector Borne Dis 44:1–11

Olayemi A, Cadar D, Magassouba N et al (2016) New hosts of the Lassa virus. Sci Rep 6:25280

Oldstone MBA (2007) A suspenseful game of "hide and seek" between virus and host. Nat Immunol 8:325–327

Oloniniyi OK, Unigwe US, Okada S et al (2018) Genetic characterization of Lassa virus strains isolated from 2012 to 2016 in southeastern Nigeria. PLoS Negl Trop Dis 12:e0006971

Olschläger S, Günther S (2012) Rapid and specific detection of Lassa virus by reverse transcription-PCR coupled with oligonucleotide array hybridization. J Clin Microbiol 50:2496–2499

Olschlager S, Lelke M, Emmerich P et al (2010) Improved detection of Lassa virus by reverse transcription-PCR targeting the 5' region of S RNA. J Clin Microbiol 48:2009–2013

Olugasa BO, Odigie EA, Lawani M, Ojo JF (2015) Development of a time-trend model for analyzing and predicting case-pattern of Lassa fever epidemics in Liberia, 2013–2017. Ann Afr Med 14:89–96

Oppliger J, Torriani G, Herrador A, Kunz S (2016) Lassa virus cell entry via dystroglycan involves an unusual pathway of macropinocytosis. J Virol 90:6412–6429

Ortiz-Riaño E, Cheng BYH, de la Torre JC, Martínez-Sobrido L (2011) The C-terminal region of lymphocytic choriomeningitis virus nucleoprotein contains distinct and segregable functional domains involved in NP-Z interaction and counteraction of the type I interferon response. J Virol 85:13038–13048

Palese P (1995) Genetic engineering of infectious negative-strand RNA viruses. Trends Microbiol 3:123–125

Pasqual G, Rojek JM, Masin M et al (2011) Old world Arenaviruses enter the host cell via the multivesicular body and depend on the endosomal sorting complex required for transport. PLoS Pathog 7:e1002232

Pasquato A, Fernandez AH, Kunz S (2018) Studies of Lassa virus cell entry. In: Salvato MS (ed) Hemorrhagic fever viruses: methods and protocols. Springer, New York, pp 135–155

Patassi AA, Landoh DE, Mebiny-Essoh Tchalla A et al (2017) Emergence of Lassa fever disease in northern Togo: report of two cases in Oti District in 2016. Case Rep Infect Dis 2017:8242313

Patterson M, Seregin A, Huang C et al (2014) Rescue of a recombinant Machupo virus from cloned cDNAs and in vivo characterization in interferon ($\alpha\beta/\gamma$) receptor double knockout mice. J Virol 88:1914–1923

Pemba CM, Kurosaki Y, Yoshikawa R et al (2019) Development of an RT-LAMP assay for the detection of Lassa viruses in southeast and south-central Nigeria. J Virol Methods 269:30–37

Perez M, Greenwald DL, de la Torre JC (2004) Myristoylation of the RING finger Z protein is essential for Arenavirus budding. J Virol 78:11443–11448

Peters CJ, Jahrling PB, Liu CT et al (1987) Experimental studies of Arenaviral hemorrhagic fevers. Arenaviruses. Springer, Berlin Heidelberg, pp 5–68

Pinschewer DD, Perez M, de la Torre JC (2005) Dual role of the lymphocytic choriomeningitis virus intergenic region in transcription termination and virus propagation. J Virol 79:4519–4526

Plowright RK, Parrish CR, McCallum H et al (2017) Pathways to zoonotic spillover. Nat Rev Microbiol 15:502–510

Polyak SJ, Zheng S, Harnish DG (1995) 5' termini of Pichinde arenavirus S RNAs and mRNAs contain nontemplated nucleotides. J Virol 69:3211–3215

Pontremoli C, Forni D, Cagliani R et al (2017) Evolutionary analysis of Old World Arenaviruses reveals a major adaptive contribution of the viral polymerase. Mol Ecol 26:5173–5188

Popkin DL, Teijaro JR, Lee AM et al (2011) Expanded potential for recombinant trisegmented lymphocytic choriomeningitis viruses: protein production, antibody production, and in vivo assessment of biological function of genes of interest. J Virol 85:7928–7932

Prescott JB, Marzi A, Safronetz D et al (2017) Immunobiology of Ebola and Lassa virus infections. Nat Rev Immunol 17:195–207

ProMED-mail (2018) Lassa fever—West Africa (32): Sierra Leone

ProMED-mail (2020) Lassa fever—West Africa (43): Liberia

Purushotham J, Lambe T, Gilbert SC (2019) Vaccine platforms for the prevention of Lassa fever. Immunol Lett. https://doi.org/10.1016/j.imlet.2019.03.008

Qi X, Lan S, Wang W et al (2010) Cap binding and immune evasion revealed by Lassa nucleoprotein structure. Nature 468:779–783

Quick J, Grubaugh ND, Pullan ST et al (2017) Multiplex PCR method for MinION and Illumina sequencing of Zika and other virus genomes directly from clinical samples. Nat Protoc 12:1261–1276

Raabe V, Koehler J (2017) laboratory Diagnosis of Lassa fever. J Clin Microbiol 55:1629–1637

Radoshitzky SR, Bào Y, Buchmeier MJ et al (2015) Past, present, and future of arenavirus taxonomy. Arch Virol 160:1851–1874

Radoshitzky S, Kuhn JH, Jahrling P, Bavari S (2018) HEMORRHAGIC FEVER-CAUSING MAMMARENAVIRUSES. In: MEDICAL ASPECTS OF BIOLOGICAL WARFARE. Borden Institute, US Army Medical Department Center and School, Health Readiness Center of Excellence

Raju R, Raju L, Hacker D et al (1990) Nontemplated bases at the 5' ends of Tacaribe virus mRNAs. Virology 174:53–59

Rang FJ, Kloosterman WP, de Ridder J (2018) From squiggle to basepair: computational approaches for improving nanopore sequencing read accuracy. Genome Biol 19:90

Rawls WE, Chan MA, Gee SR (1981) Mechanisms of persistence in arenavirus infections: a brief review. Can J Microbiol 27:568–574

Reguera J, Cusack S, Kolakofsky D (2014) Segmented negative strand RNA virus nucleoprotein structure. Curr Opin Virol 5:7–15

Reignier T, Oldenburg J, Noble B et al (2006) Receptor use by pathogenic Arenaviruses. Virology 353:111–120

Rivers TM, McNair Scott TF (1935) Meningitis in man caused by a filterable virus. Science 81:439–440

Robinson JE, Hastie KM, Cross RW et al (2016) Most neutralizing human monoclonal antibodies target novel epitopes requiring both Lassa virus glycoprotein subunits. Nat Commun 7:11544

Rodrigo WWSI, de la Torre JC, Martínez-Sobrido L (2011) Use of single-cycle infectious lymphocytic choriomeningitis virus to study hemorrhagic fever Arenaviruses. J Virol 85:1684–1695

Rojek JM, Campbell KP, Oldstone MBA, Kunz S (2007a) Old World arenavirus infection interferes with the expression of functional alpha-dystroglycan in the host cell. Mol Biol Cell 18:4493–4507

Rojek JM, Spiropoulou CF, Campbell KP, Kunz S (2007b) Old World and clade C New World arenaviruses mimic the molecular mechanism of receptor recognition used by alpha-dystroglycan's host-derived ligands. J Virol 81:5685–5695
Rojek JM, Sanchez AB, Nguyen NT et al (2008) Different mechanisms of cell entry by human-pathogenic Old World and New World Arenaviruses. J Virol 82:7677–7687
Rojek JM, Moraz M-L, Pythoud C et al (2012) Binding of Lassa virus perturbs extracellular matrix-induced signal transduction via dystroglycan. Cell Microbiol 14:1122–1134
Rowe WP, Murphy FA, Bergold GH et al (1970) Arenoviruses: proposed name for a newly defined virus group. J Virol 5:651–652
Ruigrok RWH, Crépin T, Kolakofsky D (2011) Nucleoproteins and nucleocapsids of negative-strand RNA viruses. Curr Opin Microbiol 14:504–510
Safronetz D, Lopez JE, Sogoba N et al (2010) Detection of Lassa virus, Mali. Emerg Infect Dis 16:1123–1126
Safronetz D, Strong JE, Feldmann F et al (2013) A recently isolated Lassa virus from Mali demonstrates atypical clinical disease manifestations and decreased virulence in cynomolgus macaques. J Infect Dis 207:1316–1327
Safronetz D, Mire C, Rosenke K et al (2015) A recombinant vesicular stomatitis virus-based Lassa fever vaccine protects guinea pigs and macaques against challenge with geographically and genetically distinct Lassa viruses. PLoS Negl Trop Dis 9:e0003736
Salami K, Gsell P-S, Olayinka A et al (2020) Meeting report: WHO consultation on accelerating Lassa fever vaccine development in endemic countries, Dakar, 10–11 September 2019. Vaccine. https://doi.org/10.1016/j.vaccine.2020.01.017
Salvato MS, Shimomaye EM (1989) The completed sequence of lymphocytic choriomeningitis virus reveals a unique RNA structure and a gene for a zinc finger protein. Virology 173:1–10
Salvato MS, Domi A, Guzmán-Cardozo C et al (2019) A single dose of modified vaccinia ankara expressing Lassa Virus-like particles protects mice from lethal intra-cerebral virus challenge. Pathogens 8. https://doi.org/10.3390/pathogens8030133
Sánchez AB, de la Torre JC (2006) Rescue of the prototypic Arenavirus LCMV entirely from plasmid. Virology 350:370–380
Schaeffer J, Reynard S, Carnec X et al (2019) Non-pathogenic Mopeia virus induces more robust activation of plasmacytoid dendritic cells than lassa virus. Viruses 11. https://doi.org/10.3390/v11030287
Schlie K, Maisa A, Freiberg F et al (2010a) Viral protein determinants of Lassa virus entry and release from polarized epithelial cells. J Virol 84:3178–3188
Schlie K, Maisa A, Lennartz F et al (2010b) Characterization of Lassa virus glycoprotein oligomerization and influence of cholesterol on virus replication. J Virol 84:983–992
Shaffer JG, Schieffelin JS, Gbakie M et al (2019) A medical records and data capture and management system for Lassa fever in Sierra Leone: approach, implementation, and challenges. PLoS ONE 14:e0214284
Shendure J, Balasubramanian S, Church GM et al (2017) DNA sequencing at 40: past, present and future. Nature 550:345–353
Shi M, Lin X-D, Chen X et al (2018) The evolutionary history of vertebrate RNA viruses. Nature 556:197–202
Shimojima M, Ströher U, Ebihara H et al (2012) Identification of cell surface molecules involved in dystroglycan-independent Lassa virus cell entry. J Virol 86:2067–2078
Shtanko O, Imai M, Goto H et al (2010) A role for the C terminus of Mopeia virus nucleoprotein in its incorporation into Z protein-induced virus-like particles. J Virol 84:5415–5422
Siddle KJ, Eromon P, Barnes KG et al (2018) Genomic analysis of Lassa Virus during an increase in cases in Nigeria in 2018. N Engl J Med 379:1745–1753
Stenglein MD, Jacobson ER, Chang L-W et al (2015) Widespread recombination, reassortment, and transmission of unbalanced compound viral genotypes in natural arenavirus infections. PLoS Pathog 11:e1004900
Strecker T, Eichler R, Meulen JT et al (2003) Lassa virus Z protein is a matrix protein and sufficient for the release of virus-like particles [corrected]. J Virol 77:10700–10705

Takah NF, Brangel P, Shrestha P, Peeling R (2019) Sensitivity and specificity of diagnostic tests for Lassa fever: a systematic review. BMC Infect Dis 19:647

Ter Meulen J, Koulemou K, Wittekindt T et al (1998) Detection of Lassa virus antinucleoprotein immunoglobulin G (IgG) and IgM antibodies by a simple recombinant immunoblot assay for field use. J Clin Microbiol 36:3143–3148

Trappier SG, Conaty AL, Farrar BB et al (1993) Evaluation of the polymerase chain reaction for diagnosis of Lassa virus infection. Am J Trop Med Hyg 49:214–221

Trombley AR, Wachter L, Garrison J et al (2010) Comprehensive panel of real-time TaqMan polymerase chain reaction assays for detection and absolute quantification of Filoviruses, Arenaviruses, and New World Hantaviruses. Am J Trop Med Hyg 82:954–960

Urata S, Yasuda J (2012) Molecular mechanism of arenavirus assembly and budding. Viruses 4:2049–2079

Urata S, Yun N, Pasquato A et al (2011) Antiviral activity of a small-molecule inhibitor of arenavirus glycoprotein processing by the cellular site 1 protease. J Virol 85:795–803

Urata S, Ngo N, de la Torre JC (2012) The PI3K/Akt pathway contributes to arenavirus budding. J Virol 86:4578–4585

Viana M, Mancy R, Biek R et al (2014) Assembling evidence for identifying reservoirs of infection. Trends Ecol Evol 29:270–279

Vieth S, Torda AE, Asper M et al (2004) Sequence analysis of L RNA of Lassa virus. Virology 318:153–168

Vogel D, Rosenthal M, Gogrefe N et al (2019) Biochemical characterization of the Lassa virus L protein. J Biol Chem. https://doi.org/10.1074/jbc.RA118.006973

Walker DH, Wulff H, Lange JV, Murphy FA (1975) Comparative pathology of Lassa virus infection in monkeys, Guinea-pigs, and *Mastomys natalensis*. Bull World Health Organ 52:523–534

Wang J, Danzy S, Kumar N et al (2012) Biological roles and functional mechanisms of Arenavirus Z protein in viral replication. J Virol 86:9794–9801

Wang MK-M, Ren T, Liu H et al (2018a) Critical role for cholesterol in Lassa fever virus entry identified by a novel small molecule inhibitor targeting the viral receptor LAMP1. PLoS Pathog 14:e1007322

Wang P, Liu Y, Zhang G et al (2018b) Screening and identification of Lassa virus entry inhibitors from an FDA-approved drug library. J Virol 92. https://doi.org/10.1128/JVI.00954-18

Watanabe Y, Raghwani J, Allen JD et al (2018) Structure of the Lassa virus glycan shield provides a model for immunological resistance. Proc Natl Acad Sci U S A 115:7320–7325

Welch SR, Guerrero LW, Chakrabarti AK et al (2016) Lassa and Ebola virus inhibitors identified using minigenome and recombinant virus reporter systems. Antiviral Res 136:9–18

Welch SR, Scholte FEM, Albariño CG et al (2018) The S genome segment is sufficient to maintain pathogenicity in intra-clade Lassa virus Reassortants in a Guinea pig model. Front Cell Infect Microbiol 8:240

Whitmer SLM, Strecker T, Cadar D et al (2018) New lineage of Lassa Virus, Togo, 2016. Emerg Infect Dis 24:599–602

Whitt MA (2010) Generation of VSV pseudotypes using recombinant ΔG-VSV for studies on virus entry, identification of entry inhibitors, and immune responses to vaccines. J Virol Methods 169:365–374

WHO (2018) 2018 Annual review of diseases prioritized under the R&D Blueprint

Wilda M, Lopez N, Casabona JC, Franze-Fernandez MT (2008) Mapping of the tacaribe arenavirus Z-protein binding sites on the L protein identified both amino acids within the putative polymerase domain and a region at the N terminus of L that are critically involved in binding. J Virol 82:11454–11460

Wiley MR, Fakoli L, Letizia AG et al (2019) Lassa virus circulating in Liberia: a retrospective genomic characterisation. Lancet Infect Dis. https://doi.org/10.1016/S1473-3099(19)30486-4

Willard KA, Alston JT, Acciani M, Brindley MA (2018) Identification of residues in Lassa Virus glycoprotein subunit 2 that are critical for protein function. Pathogens 8. https://doi.org/10.3390/pathogens8010001

Wohl S, Schaffner SF, Sabeti PC (2016) Genomic analysis of viral outbreaks. Annu Rev Virol 3:173–195
Wolff H, Lange JV, Webb PA (1978) Interrelationships among Arenaviruses measured by indirect immunofluorescence. Intervirology 9:344–350
Woyessa AB, Maximore L, Keller D et al (2019) Lesson learned from the investigation and response of Lassa fever outbreak, Margibi County, Liberia, 2018: case report. BMC Infect Dis 19:610
Wulff H, Johnson KM (1979) Immunoglobulin M and G responses measured by immunofluorescence in patients with Lassa or Marburg virus infections. Bull World Health Organ 57:631–635
Wylie TN, Wylie KM, Herter BN, Storch GA (2015) Enhanced virome sequencing using targeted sequence capture. Genome Res 25:1910–1920
Ye SH, Siddle KJ, Park DJ, Sabeti PC (2019) Benchmarking metagenomics tools for taxonomic classification. Cell 178:779–794
Young PR, Howard CR (1983) Fine structure analysis of Pichinde virus nucleocapsids. J Gen Virol 64(Pt 4):833–842
Zapata JC, Poonia B, Bryant J et al (2013) An attenuated Lassa vaccine in SIV-infected rhesus macaques does not persist or cause arenavirus disease but does elicit Lassa virus-specific immunity. Virol J 10:52
Zaza AD, Herbreteau CH, Peyrefitte CN (2018) Description and characterization of a novel live-attenuated tri-segmented Machupo virus in Guinea pigs. Virol J 15:99
Zhang X, Yan F, Tang K et al (2019) Identification of a clinical compound losmapimod that blocks Lassa virus entry. Antiviral Res 167:68–77
(2019a) Aldatu biosciences expands PANDAA platform beyond HIV with SBIR for the first pan-lineage Lassa fever test. In: Aldatu biosciences. https://www.aldatubio.com/news/aldatu-biosciences-expands-pandaa-platform-beyond-hiv-with-sbir-for-the-first-pan-lineage-lassa-fever-test/. Accessed 18 Jan 2020
(2019b) ProMED-mail post. https://www.promedmail.org/post/6653653. Accessed 19 Sept 2019
(2017) WHO|Lassa fever—Benin, Togo and Burkina Faso. https://www.who.int/csr/don/10-march-2017-lassa-fever-benin-togo-burkina-faso/en/
(2016) WHO|Lassa fever—Benin. https://www.who.int/csr/don/13-june-2016-lassa-fever-benin/en/
(2018) WHO|Lassa fever R&D. https://www.who.int/blueprint/priority-diseases/key-action/LassaVirusVaccineTPP.PDF?ua=1

Ecology of Lassa Virus

Allison R. Smither and Antoinette R. Bell-Kareem

Contents

Abstract Individuals living in endemic hotspots of Lassa fever have recurrent exposure to Lassa virus (LASV) via spillover from the primary host reservoir *Mastomys natalensis*. Despite *M. natalensis* being broadly distributed across sub-Saharan Africa, Lassa fever is only found in West Africa. In recent years, new

A. R. Smither (✉) · A. R. Bell-Kareem
Department of Microbiology and Immunology, Tulane University School of Medicine, New Orleans, LA, USA
e-mail: asmither@tulane.edu

Current Topics in Microbiology and Immunology (2023) 440: 67–86
https://doi.org/10.1007/82_2020_231

Published Online: 11 February 2021

LASV reservoirs have been identified. Erudition of rodent habitats, reproduction and fecundity, movement patterns, and spatial preferences are essential to institute preventative measures against Lassa fever. Evolutionary insights have also added to our knowledge of closely related mammarenavirus distribution amongst rodents throughout the continent.

1 Introduction

Lassa virus (LASV) causes Lassa fever, a severe viral hemorrhagic fever spread by rodents in West Africa. With an estimated 100,000–300,000 cases and more than 5000 deaths yearly, it is the most consequential rodent-borne virus worldwide (Monath 1975). Individuals contract LASV through inhalation of aerosolized urine in droplets or dust particles, as well as via ingestion of food or water or contact with fomites contaminated with rodent urine or droppings. Rodents are nocturnal animals that source food during the night; open food or water containers and other fomites can become contaminated with rodent excrement (Leirs 1994). Upon inhalation, consuming or handling the infectious material, the virus can enter through contact with mucous membranes, cuts in the skin, or orally, where it infects tissue resident macrophages (Kenyon et al. 1992; Walker et al. 1982). For reasons not yet fully understood, the infection will either be contained or disseminate throughout the body to cause fulminant Lassa fever. With no vaccine or other specific treatment measures available at the time of publication, rodent exclusion methods are currently the most effective preventative measure against Lassa fever.

Lassa fever and LASV were first identified in 1969, prompting a search for the source of the then novel virus. Efforts to locate LASV in tissues isolated from rodents or hedgehogs near patients' homes in Jos, Nigeria, were unfruitful (Carey et al. 1972). Unlike earlier outbreaks, the 1972 outbreak of Lassa fever in the Eastern Province of Sierra Leone saw cases stemming from individual communities rather than from nosocomial infections, further suggesting into the possibility of a non-human reservoir (Fraser et al. 1974). The subsequent findings from this initial study determined that the peri-domestic rodent *Mastomys natalensis* was the major if not sole reservoir for LASV. Since then, ensuing cases and outbreaks of Lassa fever have been extensively monitored over time alongside further exploration into the ecology of *M. natalensis* and other rodents and their interactions with LASV.

Critical to the understanding of Lassa fever epidemiology is the natural ecology of the rodent reservoir *M. natalensis*. Of hundreds of rodent species present throughout sub-Saharan Africa, *M. natalensis* is the most broadly distributed (Wilson and Reeder 1993). It is a serious agricultural pest (Mulungu et al. 2013; Makundi et al. 2008; Mariën et al. 2018). However, contracting Lassa fever from rodents is more likely to occur when the rodents are driven inside homes, as opposed to infection in the fields (Mariën et al. 2018). Indeed, in homes where more *M. natalensis* are present higher a seroprevalence of LASV is seen. This trend has been repeatedly observed, as pointed focal distribution of *M. natalensis* mirrors

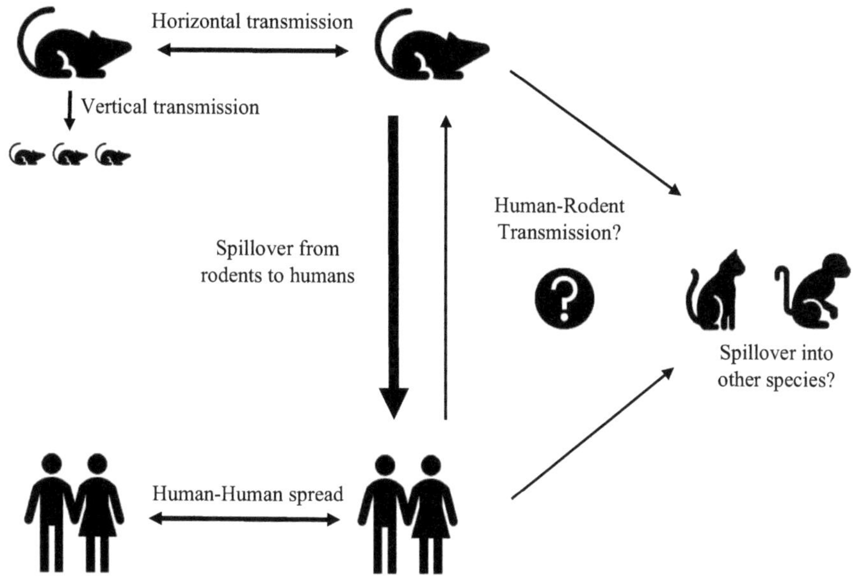

Fig. 1 Overview of transmission routes of LASV. Rodents spread the virus amongst themselves through both horizontal and vertical transmission. Most human cases of Lassa fever derive from contact with rodent excrement. The remaining cases arises from direct or indirect contact with infected individuals. It is unknown whether other species come into contact with LASV via infected rodents or whether humans have the ability to transfer the virus back to rodents or other species

LASV seroprevalence (Demby et al. 2001; Lecompte et al. 2006; Mccormick et al. 1987; Monath et al. 1974) (Fig. 1).

Here, we discuss the ecology of Lassa fever in *M. natalensis*, the identification of additional reservoirs, and the capacity for LASV spillover to other species. Epizoology studies undertaken throughout West Africa indicate high levels (upward of 30%) of LASV presence in *M. natalensis*. Seasonality and fecundity of *M. natalensis* with respect to the temporal dynamics of LASV in rodents provide insight into human Lassa fever, as well as strategies for Lassa fever prevention. Finally, we discuss the relatively narrow distribution of LASV in comparison with other Old-World mammarenaviruses found in *M. natalensis* and other related rodents throughout sub-Saharan Africa.

2 Reservoirs of Lassa Virus

2.1 *Mastomys Natalensis*

After the initially detected outbreaks of Lassa fever occurred in Nigeria and Sierra Leone, electron microscopy determined that LASV was structurally similar to lymphocytic choriomeningitis virus (LCMV). As LCMV is found worldwide except

Antarctica in the common house mouse *Mus musculus domesticus*, attempts to isolate LASV from African rodents were undertaken (Arata and Gratz 1975). Early attempts at rodent trapping to isolate LASV were unsuccessful, possibly due to the small survey size and scope of animals collected (Monath 1975). During the 1972 outbreak in Sierra Leone, a thorough search of local vertebrate animals was conducted in two of the villages where the cases originated (Monath et al. 1974). Snap-frozen organs from collected animals, predominantly rodents and bats, were crushed, and the extracted supernatant was used to test for potential cytopathic effect (CPE). The only specimens testing positive for CPE were from *Mastomys natalensis*. This finding was later confirmed during a subsequent outbreak investigation of Nigerian rodents where *M. natalensis* was the predominant species where CPE was observed from organ extracts of collected animals (Wulff et al. 1975). Multiple studies have concluded that *M. natalensis* is the predominant reservoir of LASV throughout its entire geographic distribution in West Africa (Lecompte et al. 2006; Leski et al. 2015; Safronetz et al. 2013; Kouadio et al. 2015).

2.2 Other Rodents

Anti-LASV IgG antibodies have been detected in multiple other rodent species: *Praomys* spp., *Mus* spp., *Rattus rattus*, and *Lemniscomys striatus*, providing evidence of occasional spillover and transient infection (Demby et al. 2001; Fichet-Calvet et al. 2014; Olayemi et al. 2016a). New evidence suggests three other species of rodents could be potential hosts to LASV. LASV has been isolated in multiple specimens of *Hylomyscus pamfi* in Nigeria and *Mastomys erythroleucus* in both Nigeria and Guinea over a period of several years by qRT-PCR (Olayemi et al. 2016b; Agbonlahor et al. 2017). In villages where the virus is isolated from *M. erythroleucus*, virus could not be obtained from *M. natalensis*. LASV and LASV-like viruses have been isolated in Pygmy mice (*Mus baoulei*) in Benin and Ghana (Kronmann et al. 2013; Yadouleton et al. 2019). These newly isolated viral sequences from "alternative" reservoirs do not cluster with viruses in the traditional four/five lineages, revealing a potentially new lineage of LASV previously unseen in circulation. Whether these species have always been reservoirs to LASV or if viral mutations allow for host-switching and persistence in a new population is unknown.

It is plausible that *M. erythroleucus* is not a newly established host reservoir of LASV. Members of the *Mastomys* genus are typically indistinguishable by morphological characteristics and thus can only be confirmed to the species level by molecular genotyping. The distinct species of *Mastomys* were initially characterized by chromosomal karyotyping or hemoglobin electrophoresis, whereas the foundational reports of LASV in *Mastomys* categorized all *Mastomys* caught as *M. natalensis* (Monath 1975; McCormick et al. 1987; Robbins et al. 1983). It will be important to appropriately identify all potential reservoirs to the species level to elucidate the potential distribution of LASV hosts and determine specific ecological niches specific to each potential host.

2.3 *Additional Hosts*

Macaques and other non-human primates (NHPs) simulate human infection upon inoculation with LASV; nearly all species of NHP used in laboratory testing manifest fatal pathology even with low titers of LASV (Malhotra et al. 2013). Interestingly, IgG antibodies to LASV have been detected in otherwise healthy Mona monkeys (*Cercopithecus mona*) and a single Anubis baboon (*Papio anubis*) in two different Nigerian states (Ogunro et al. 2019). A single Mona monkey was also found to be both antibody and antigen positive for LASV, though none of the animals were positive for IgM. Virus was unable to be detected by RT-PCR in this animal. Nearly all the IgG positive NHPs were found scavenging refused near the Lagos lagoon, though the Lagos state has no previously reported cases or sero-prevalence of LASV. Exactly how NHPs encountered LASV or a LASV-like virus, as well as their role, if any, in the sylvatic transmission cycle remains to be determined. To date, no other vectors (ticks, mosquitos, etc.) have been identified as a source or intermediate of LASV. However, a recent study demonstrated IgG presence in insectivorous shrews (*Crocidura* spp.) (Olayemi et al. 2018). Sequencing and sero-epidemiological studies focusing on non-rodent mammals will be needed to better elucidate the diversity of reservoir and spillover species.

3 Epizoology Studies of Lassa Virus

3.1 *Guinea*

Much of the research about the spatial and temporal dynamics of LASV in rodents has been conducted in Guinea. Nationwide trapping studies surveying every eco-zone were conducted between 1996–1997 and 2002–2005, with additional ad hoc trapping in endemic hot spots along the Guinea–Sierra Leone border and in the eastern half of the country. Antibodies to LASV are found in approximately 10% of *M. natalensis* nationwide, with up to 27% seroprevalence seen in villages in high prevalence regions (Demby et al. 2001; Fichet-Calvet et al. 2014). In spite of high IgG antibody levels, only 5% of *M. natalensis* had measurable antigen levels (Demby et al. 2001). LASV has been detected via RT-PCR in 10–15% of *M. natalensis*, with higher focality in certain endemic villages (Lecompte et al. 2006; Fichet-Calvet et al. 2007; Lalis et al. 2013). Most specimens were not antibody and virus/antigen positive at the same time, suggesting immunological clearance of the virus. LASV in *M. erythroleucus* was detected in 6/14 specimens in coastal Guinea (Mandia Oula), indicating a new potential host of the virus (Olayemi et al. 2016b). No *M. natalensis* were caught in the village where the sampling occurred. Humans in this same village had a 35% seroprevalence rate of LASV despite little other evidence of LASV in the region (Lukashevich et al. 1993). As the virus clusters phylogenetically with Lineage IV Josiah as seen in Sierra Leone and forest Guinea, it is likely that the virus was imported to the area and underwent host-switching to adapt to its new environment (Olayemi et al. 2016b).

3.2 *Sierra Leone and Liberia*

LASV was first isolated in cell culture from *Mastomys natalensis* captured in the eastern province of Sierra Leone in 1972 during the first documented non-nosocomial outbreak of Lassa fever (Monath et al. 1974). This study showed that higher presence of *M. natalensis* corresponded with higher LASV seroprevalence in the human population. Intriguingly, rodent trapping in Zorzor, Liberia, during the 1972 outbreak did not yield any positive specimens (Monath 1975); subsequent rodent trapping studies have not been performed to date. Ensuing rodent studies performed in Sierra Leone have focused on the anthropological issues surrounding Lassa fever.

Two of the only epizoology studies in Sierra Leone took place in the 1980s. In the first investigation, case–control trapping found viremic *M. natalensis* were ten times as likely to be found in a patient's house than in a control house (Keenlyside et al. 1983). A third of all *M. natalensis* collected in case houses had evidence of LASV viremia. However, all houses sampled had similar seroconversion rates to LASV, indicating potential movement of rodents from other non-trapped houses nearby to continue the chain of infection. The same study found higher antibody prevalence in individuals from households with a Lassa fever case than without (30 vs. 20%).

A second study examined LASV seroprevalence in villages across Sierra Leone (8–52% depending on the village and region) with rodent LASV viremia (0–81%; overall 11.2%), though no distinct correlation was seen with rodent viral presence and human seroprevalence (McCormick et al. 1987). LASV antibodies occurred in up to approximately 35% of all *M. natalensis*; concurrent viremia and antibody detection were found in about 5% of animals. No LASV was observed in *M. natalensis* located along the Atlantic coast of the southern province of Sierra Leone, despite serological evidence of LASV infection in humans, raising the possibility that there is an additional reservoir of LASV in this area. Within the northern and eastern provinces, two different species of *Mastomys* were seen. One species predominated in the northern plateau, an area that had lower seroprevalence and seroconversion rates than in the eastern province where the other species dominated. The two different Mastomys species were delineated by karyotyping, as genotyping at the time was not able to distinguish two separate species. With advances in genotyping and PCR techniques, a new undertaking of LASV distribution in rodents in the eastern province isolated virus from 20% of captured *M. natalensis* (Leski et al. 2015). Within the small study range, high sequence diversity within Lineage IV was seen, demonstrating local clustering of the virus within specific villages.

3.3 Nigeria

LASV was first successfully extracted from 8/151 rodents (5 *M. natalensis*, 2 *Mus musculus*, and 1 *Rattus rattus*) in the Benue-Plateau and North-Eastern States in 1975 (Wulff et al. 1975). Since then, surveys have been orchestrated throughout Nigeria to measure LASV in the rodent population. In one survey conducted in Ekpoma, Edo State, 25% of specimens collected were *M. natalensis*, of which nearly half were reactive to LASV complement-fixing antibody (Okoror et al. 2005). An extraordinarily high presence of LASV in the rodent population would explain why Lassa fever is endemic in the Edo State. Other studies in the Edo State have shown lower prevalence in LASV in the rodent population. One study found only 16/274 *M. natalensis* positive for LASV, and another could not isolate viral sequences in any *M. natalensis* found (Olayemi et al. 2016a, b). All LASV isolates clustered to Lineage II.

A more extensive collection of 1500 specimens in Edo, Delta, and Bayelsa states found *M. natalensis* comprised only 2% of rodent species collected (Agbonlahor et al. 2017). All but two the *M. natalensis* found in Edo and half found in Delta were LASV positive, with only one positive *M. natalensis* found in the non-endemic state of Bayelsa. The most recent and extensive trapping studies show a wide range of LASV or LASV-like virus seroprevalence in multiple rodent species within LASV endemic areas, ranging anywhere from 1 to 52% depending on the locality (Olayemi et al. 2016b,2018). This evidence suggested that the rodent hosts of LASV may have expanded in Nigeria. It should also be noted that investigations that did not examine other species for LASV in Nigeria may not have uncovered the pervasiveness of the virus in rodent populations.

3.4 Other West African Countries

Though not traditionally considered endemic for Lassa fever, several West African countries in addition to the Mano River Union countries (Guinea, Sierra Leone, Liberia) and Nigeria have reported cases of Lassa fever over the past 10–15 years. Côte d'Ivoire has not experienced a confirmed case of Lassa fever during the prodromal phase, despite LASV circulating in the rodent population (Akoua-Koffi et al. 2006; Emmerich et al. 2008). Extensive trapping surveys conducted between 2003 and 2005 found serological evidence LASV or LASV-like virus in four different species of rodent including *M. natalensis* (Coulibaly-N'Golo et al. 2011). Two novel mammarenaviruses were also found in the southwest part of the country in *Hylomyscus* spp. and *Mus (Nannomys) setulosus*, respectively. The virus found in *Hylomyscus* spp. (named Gbagroube) closely resembled LASV Lineage II, while the other virus (Menekre) clustered more with East African mammarenaviruses Mopeia and Mobala. In 2013, LASV RNA from *M. natalensis* was isolated and sequenced in 2013; these strains converged with strain AV, which was the first strain of LASV detected from a traveler in the area (Kouadio et al. 2015; Manning et al. 2015). Epizoology studies in Mali found an overall prevalence rate of LASV across all ecozones to be 6.8%. Prevalence rates of LASV in rodents in the southern region hover near 20%, with individual villages ranging from 0 to 52% (Safronetz et al. 2013). LASV in Mali clusters with lineage AV.

The first human serosurveys in Ghana indicated the overall LASV seroprevalence to be 5.2%, with individual villages as high as 16% (Nimo-Paintsil et al. 2019). Rodents screened in the villages also showed no evidence of LASV via RT-PCR in *M. natalensis,* though two novel mammarenaviruses were discovered in two separate *Mus* species (Kronmann et al. 2013). The first virus, found in the central of the country near the border with Togo, clustered with Lineage I (Pinneo). The second, discovered in the northwest, grouped with LCMV. With clear serological evidence of LASV in the population, it is plausible that virus either has a different reservoir in Ghana or that *M. natalensis* has circulating levels beneath the limit of detection with conventional PCR techniques. Despite the discovery of two novel mammarenaviruses near the borders of Togo and Burkina Faso, no rodent trapping studies designed to detect LASV have been performed in these countries to date. In Benin, which borders Nigeria to the east, three pygmy mice (*Mus baoulei*) captured in central part of the country were PCR positive for LASV, clustering with strain AV (Yadouleton et al. 2019).

Senegal has had no confirmed cases of Lassa fever. A potential case of Lassa fever in eastern Senegal was reported in the early 1970s, but no evidence of LASV was found in human or rodent sera (Monath 1975). Rodent trapping performed in the 1980s revealed a very low overall LASV seroprevalence rate of about 1.2%, jumping up to 2.1% in *M. natalensis* (Saluzzo et al. 1988). Likewise, while the Central African Republic has never reported a case of Lassa fever, related Old-World mammarenaviruses have been isolated from *Arvicanthus* spp. and *Praomys* spp., which have been demonstrated to cross-react with LASV and other related mammarenaviruses (Gonzalez et al. 1983; Swanepoel et al. 1985). This cross-reactivity calls into question the specificity of the few seroprevalence studies undertaken in the rodent population.

4 Dynamics of Lassa Virus in *Mastomys natalensis*

4.1 Spatial and Temporal Dynamics

The ecology of Lassa fever is entwined with the environmental, geographic, and spatial factors affecting the distribution of *M. natalensis*. *Mastomys* biome preference has been inferred from studies conducted in Guinea (West Africa) and Tanzania (East Africa). *M. natalensis* density is highest in tropical savanna regions in both West and East Africa. West Africa also has a high density of *M. natalensis* in forested regions, while forested areas of East Africa have lower *M. natalensis* density (Makundi et al. 2008; Demby et al. 2001; Lecompte et al. 2006; Borremans et al. 2011). *Rattus* and *Mus* species tend to predominate over *M. natalensis* in coastal and urban areas (Demby et al. 2001; Keenlyside et al. 1983). Higher human LASV seroprevalence is found in forested and savannah regions, mirroring general *M. natalensis* distribution in West Africa (Lukashevich et al. 1993). An explanation for the dearth of *M. natalensis* along the coast could be competition by imported rodents. It is plausible that European-lineage rodents such as *Rattus* and *Mus* may have been brought over on ships during slave transit and trade expeditions where they settled in the nearby shores and outcompeted other African-lineage rodents in environments that matched their ideal niche. Preferences in building construction in relation to rodent burrowing preferences (e.g., concrete vs. mud floors; metal vs. thatched roofs) could explain an absence of *M. natalensis* in urban areas (Bonwitt et al. 2017). These differences in building furnishes and construction in rural vs. urban housing allow for increased rodent burrowing and hiding, contributing to infestations with potentially infected animals in rural areas (Bonwitt et al. 2017; Bonner et al. 2007).

At the village level, *M. natalensis* is preferentially found inside homes as opposed to open fields (Mariën et al. 2018; Fichet-Calvet et al. 2007; Demartini et al. 1975), especially during the dry season. *M. natalensis* population levels decrease from the outskirts of a village to its core, potentially due to lack of nearby vegetation in adjacent gardens or forest (Demby et al. 2001; Fichet-Calvet et al. 2007). During the rainy season, food and resources are abundant outside of homes in proximal fields. The burning and clearing of fields during the dry season can lead to aerosolization of excrement and driving mice to enter homes in search of food, all activities that increase the likelihood of individuals encountering an infected rodent. The prevalence of LASV in *M. natalensis* remains constant year-round in homes, which seemingly contradicts the cyclical nature of Lassa fever (Fichet-Calvet et al. 2007). The authors of this study suggested the possibility of non-seasonal breeding inside homes in which the mice breed year-round (as opposed to a peak reproductive period at the beginning of the rainy season), contributing to consistent viral levels within the rodents. When other rodents such as *Mus* spp. and *Rattus* spp. out-compete *M. natalensis* in homes, reduced human seroprevalence and rodent viral presence are seen (Demby et al. 2001; Monath et al. 1974). Furthermore, infectious rodents tend to be clustered in a select few houses as opposed to evenly distributed across homes in a village (Demby et al. 2001; Mariën et al. 2020). The in-home LASV prevalence rate should be further investigated to better understand the temporal dynamics of Lassa fever.

Lassa fever appears to fluctuate seasonally. Most Lassa fever cases occur during the West African dry season spanning roughly from November to April, peaking in January (Bausch et al. 2001; Fichet-Calvet et al. 2008). Statistical modeling has shown regions in West Africa most at risk of Lassa fever epidemics receive heavy amounts of rain, up to 3000 mm during the rainy season spanning from May–October. This correlates with observational studies of *M. natalensis* reproduction and fecundity in Guinea showing robust turnover in the rodent population from the start of the rainy season to its end, possibly an effect of stabilized food and water resources (Fichet-Calvet et al. 2008). Similar results have been observed in Tanzania, where *M. natalensis* population levels peak during the biannual rainy seasons (Mulungu et al. 2013; Borremans et al. 2011). There is conflicting evidence of overall LASV presence in *M. natalensis* in the rainy season or during the dry season (Demby et al. 2001; Fichet-Calvet et al. 2007). Other mammarenaviruses, such as LCMV and Junín virus, have increased viral presence during the breeding season (Tagliapietra et al. 2009; Mills et al. 1992). The increase in virus during the breeding season is highly indicative of vertical transmission as will be discussed below.

4.2 Transmission Modalities of Lassa Virus

The extent of horizontal and vertical transmission of LASV and other closely related OW mammarenaviruses in the host reservoir has been hotly debated. Both horizontal and vertical transmissions have been demonstrated in the laboratory. LASV-inoculated *Mus musculus* neonates develop low-level persistent infections, while inoculated adults develop transient infections. In this study, both neonates and adults developed antibodies to the viruses shortly upon inoculation, which remained for the duration of the study (Walker et al. 1975; Borremans et al. 2015). Colonies of mice (*M. musculus*) persistently infected with LCMV via vertical transmission have been established (Sun et al. 2014; Kunz et al. 2006) where they develop lifelong viremia yet have low antibody titers against the virus (Oldstone 2002). While the LCMV model may parallel LASV infection in *M. natalensis*, this feature has not been experimentally validated. LASV has been found in placental and fetal tissues as well as breast milk of humans, further suggesting that vertical transmission may occur as part of the transmission cycle outside of the reservoir (Theiler et al. 2009). Observations of wild-caught animals have noted that rodents are infrequently virus and antibody positive at the same time, pointing to vertical transmission (Demby et al. 2001; Borremans et al. 2011; Mariën et al. 2017). In transiently infected animals, the virus is able to persist in secretions for months at a time (Walker et al. 1975). Aerosols generated from mucous membranes are stable for short periods of time and may be a key component of intra-rodent horizontal transmission when rodents engage in fighting, nosing, grooming, and other social activities (Stephenson et al. 1984; Vanden Broecke et al. 2018). From an evolutionary context, vertical transmission provides the most logical explanation of viral persistence over generations. Although laboratory data provides a solid foundation to build hypotheses upon, discrepancies are seen between laboratory and natural conditions and must be taken into consideration when drawing conclusions.

The longitudinal transmission dynamics of OW mammarenaviruses in *M. natalensis* remains understudied. During an evaluation of rodents from LASV endemic Guinean villages, it was determined that older mice have higher LASV seroprevalence than young and juvenile (adolescent) mice, indicative of horizontal transmission (Demby et al. 2001; Fichet-Calvet et al. 2014). It was also revealed that LASV seroprevalence was lowest in juvenile mice and higher in young (infant) mice, indicating a potential transfer of protective antibodies from mother to fetus in utero or through breast milk. At the same time, LASV has been detected in young mice in other studies, which could be an indicator of vertical transmission (Demby et al. 2001; Fichet-Calvet et al. 2008). A study of related Mopeia virus revealed peak viral RNA presence in young rodents, indicative of either vertical transmission or horizontal transmission preferentially occurring in younger rodents (Borremans et al. 2011). This virus may not necessarily be persistent for the life of the animal in nature. Taken together, it is probable that an equal combination of vertical and horizontal transmission enables continual persistence in the LASV host populations.

Assuming undetectable but still extant virus in the maternal population, LASV presence peaks and a healthy amount of antiviral antibody is observed during the rainy season where there is high turnover within the *M. natalensis* population. During this time, the virus is spread via vertical transmission to newborn mice, or via horizontal transmission to mice that have not received in utero antibodies. As transient maternal antibodies protecting young mice from infection wane as they mature during the dry season, a recrudescence of horizontally transmitted virus occurs from other animals in the population, or perhaps "reactivating" the virus within the now juvenile rodents. Human activity and habitat changes bring the mice in closer contact with humans as indicated by the spike of human cases of Lassa fever. As the dry season draws to a close and the rainy season begins, nearly all mice will have been exposed the virus and viral levels drop once again until new litters of mice born and the cycle repeats. It is important to note that *M. natalensis* is capable of breeding year-round which could account for cases seen during the rainy season.

4.3 Stopping the Cycle of Transmission

Certain behaviors may promote spread of the virus from rodents to humans. Consumption of rodent bushmeat, at times the sole source of protein in rural areas, is associated with higher prevalence of antibodies to Lassa compared to communities where rodents are not eaten (Meulen et al. 1996). Indeed, a recent case of Lassa fever in Benin can be traced back to regular consumption of rodent bushmeat by the patient (Attinsounon et al. 2018). Activities such as agriculture, logging, and mining can contribute to forest clearing and thus increase rodent–human interactions in a more "ideal" environment for rodents, promoting conditions ripe for spillover. Much of the economies of the high Lassa fever countries of Guinea, Sierra Leone, and Liberia are dependent on mineral mining and subsistence agriculture (Central Intelligence Agency. The World Factbook 2019). In these areas, education and proper rodent control is vital to prevent the further spread of disease.

There are multiple methods of rodent control used throughout West Africa. Both trapping rodents and dispersing rodent poison are effective yet expensive, making it difficult to implement long term without coordinated efforts from communities, local health authorities, and government backing (Mari Saez et al. 2018). The most sustainable rodent control measures are preventing rodents entering homes and further rodent proofing all food sources. Cats and dogs are being explored as rodent control agents, though they may bring their own public health issues (Haun et al. 2019). In the absence of a vaccine, rodent exclusion measures are necessary to protect those at risk (Sogoba et al. 2012; Olayiwola and Bakarey 2017; Khan et al. 2008; Roberts 2018).

5 Evolutionary History of Old-World Mammarenaviruses and Their Rodent Hosts

5.1 *The Distinct Niche of Lassa Virus in West Africa*

Members of the *Mastomys* genus must be genotyped to delineate the exact species present. Upon the widespread adaptation of PCR in laboratories, "DNA barcoding" using mitochondrial DNA (mtDNA) microsatellite genes has been used to aid phylogenetic inference of different rodent species and subspecies (Lecompte et al. 2008,2002). Mitochondrial DNA is particularly useful for evolutionary and phylogeny studies as it mutates faster than the rest of the nuclear genome, enabling minute differences over relatively short periods of time to be quantified and mapped (Parson et al. 2000; Brown et al. 1979). Phylogenetic analysis using the mtDNA gene *cytochrome b* has distinguished eight different *Mastomys* species, of which only *M. natalensis* and *M. erythroleucus* have been associated with Old-World (OW) mammarenaviruses (Olayemi et al. 2016b).

In-depth microsatellite analysis of *M. natalensis* mtDNA has also been employed to further characterize differences between members of this species and its subspecies throughout Africa (Colangelo et al. 2013; Gryseels et al. 2017). This insight has shown that *M. natalensis* has different clades and sub-clades that vary slightly throughout the continent. The clades and sub-clades are generally constrained to a specific geographic area separated by natural boundaries such as lakes, mountains, and rivers. Clade A is predominantly found in West Africa entirely west of the Congo River and a small sub-clade found around Lake Victoria in Kenya. Clade B is restricted to Eastern Africa, bound by Great Rift Valley and Eastern Arc Mountains in Tanzania to the west and the Kalahari Basin and Orange River in the south. Since the discovery of LASV, several OW mammarenaviruses have been found in different *M. natalensis* sub-clades across Africa. Only sub-clade A1 in West Africa has been associated with LASV; other closely related mammarenaviruses seem to be associated with its own particular sub-clade (Olayemi et al. 2016a; Gryseels et al. 2017). The sub-clades of *M. natalensis* may explain the spatial spread of highly related mammarenaviruses throughout Africa.

Even within the LASV-specific sub-clade, certain populations of rodents are more likely to carry the virus than others. There is no evidence of *M. natalensis* migrating long distances along trade routes and major thoroughfares, nor even short distances outside of their commensal human habitats (Lalis and Wirth 2018). To establish a population, these rodents would have had to come along in food and storage containers during periods of human migration and re-settle in new areas. As such, LASV-sustaining populations of *M. natalensis* are seen within small geographic and inquiline niches, as evidenced by the low rates of gene flow and high co-sanguinity amongst LASV-positive animals (Lalis et al. 2013, 2015). The dependency on human transport to spread the rodents (and thus the virus) may explain why there is inter-village variability within a localized geographic area.

5.2 *Unraveling the Co-speciation Hypothesis*

In addition to the mammarenaviruses found exclusively in *M. natalensis*, other mammarenaviruses in closely related rodents have also been discovered (reviewed in Gryseels et al. 2015); see also Fig. 2. All rodents are either members of the same tribe *Praomyini* or a species of *Mus* (Lecompte et al. 2008). This widespread distribution leads to the original theory of co-speciation, where an individual rodent species becomes the host to a specific mammarenavirus over millions of years of evolution (Gonzalez et al. 1986; Bowen et al. 1997). Overlaying molecular clock analyses of OW mammarenaviruses with African murine radiation and the respective phylogenetic trees show considerable overlap (Bowen et al. 1997). It was believed that a common LCMV-like arenavirus progenitor may have infected a common ancestor prior to its split into different muroid subfamilies. From there, muroids began to spread from Asia to Europe, Africa, and the Americas and subsequently evolved (Lecompte et al. 2008). According to this hypothesis, OW mammarenaviruses arose from the *Murinae* subfamily and New-World (NW) mammarenaviruses arose from the *Sigmodontinae* subfamily around 30 million years ago (Arata and Gratz 1975; Gonzalez and Duplantier 1998). Over time, mammarenaviruses evolved with specific rodent species: LCMV with *Mus* spp., NW mammarenaviruses with *Calomys* and *Neotoma spp*, and OW mammarenaviruses with the *Stenocephalomys* complex (specifically *Mastomys, Praomys,* and *Hylomyscus* spp.) (Lecompte et al. 2002; Schenk et al. 2013). Fossil evidence indicates that the *Mastomys* genus arose during the Late Pliocene era approximately 2.5–3.5 million years ago in migration from Eurasia to Africa. This evolutionary event allowed for further co-speciation between specific rodent species and mammarenaviruses to develop throughout Africa (Lecompte et al. 2008). Recent reports of additional OW mammarenaviruses in brown rats (*Rattus norvegicus*) in China, Cambodia, and Thailand fit the model of co-speciation from a common viral ancestor, yet also showed an expanded range of potential *Murinae* hosts beyond the *Praomyini* tribe (Li et al. 2015; Blasdell et al. 2016).

A thorough analysis of the ribosomal-dependent RNA polymerase of mammarenaviruses reveals the division between OW and NW mammarenaviruses occurred as recently 45,000 years ago (ya), with OW mammarenaviruses

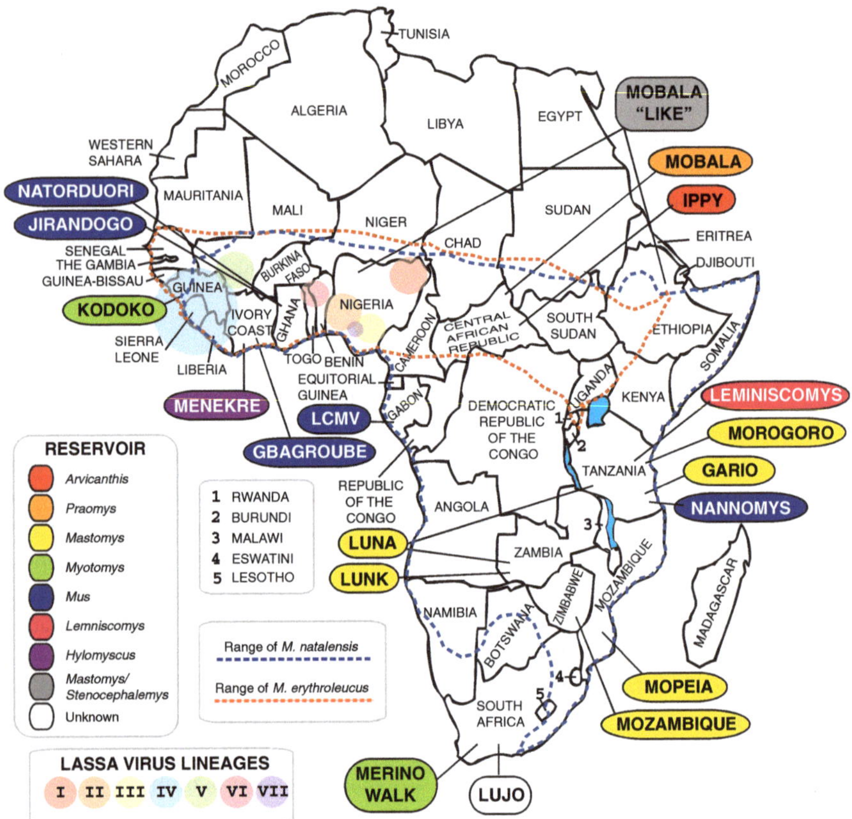

Fig. 2 Location and reservoir of known Old-World mammarenaviruses throughout sub-Saharan Africa. While LASV is exclusively found in West Africa (colored circles), the approximate range of *M. natalensis* (blue dashed line) and *M. erythroleucus* (orange dashed line) is shown to demonstrate the potential for expanded geographic distribution of LASV. Recently, it has been proposed that some of the novel mammarenaviruses in West Africa may be new lineages of LASV; this would greatly expand the number of potential reservoirs of LASV. Mammarenaviruses found in East Africa form a distinct clade within Old-World mammarenaviruses

originating between 23,000 and 1800 ya (Forni et al. 2018). Such a short evolutionary history cannot not be explained by co-divergence. In this vain, deep sequencing of clinical isolates of LASV from Sierra Leone and Nigeria indicates the evolutionary origin of LASV in Nigeria 1000–2000 years ago before spreading across West Africa indicating further disconnect between speciation and viral evolution (Andersen et al. 2015). It is thought that colonial rule, common travel, and slave trade may have allowed LASV to spread westward from Nigeria to other regions of West Africa around 150–200 years ago (Lalis et al. 2013; Manning et al. 2015).

A persistent flaw of the co-speciation model has been revealed by the presence of NW Tacaribe virus in phyllostomid bats, mosquitos, and ticks in the Caribbean, demonstrating a wide taxonomic difference between potential mammarenavirus hosts (Downs et al. 1963; Sayler et al. 2014). The recent discovery of arenaviruses in reptiles and fish further proves the broad diversity of *Arenavirus* hosts. Indeed, this has since led to reclassification of *Arenaviridae* from a single *Arenavirus* genus into four different geneses: Mammarenaviruses encompassing OW and NW arenaviruses primarily found in rodents, *Reptarenavirus* and *Hartmanivirus* in reptiles, and *Antennavirus* in fish (Shi et al. 2018; Arenaviridae 2018). Subsequently, the *Arenaviridae* family has since merged with members of the now-defunct *Bunyaviridae* family to form a new order *Bunyavirales* (Arenaviridae 2018). This taxonomic change better reflects the common genome and structural similarities between family members of this order. Genetic reassortment and recombination between these related viruses may have allowed for sylvatic transmission between species, allowing a novel host niche to evolve, emerge, and further diverge over short periods of time.

6 Conclusions

Lassa fever has a multifactorial etiology intricately woven between man and nature. Seasons, geography, age, and various human and rodent factors determine whether one will acquire an LASV infection. LASV will continue to evolve over time, potentially creating more new reservoir hosts or changes in pathogenicity. Climate change and increasing globalization will also likely enlarge the Lassa fever zone across West African and perhaps even across continental borders. An expansion bringing Lassa fever into larger population centers could result in a major public health emergency. Community hygiene practices, along with expanded surveillance and observations of both humans and rodents, are needed to trace outbreaks and prevent them before they start. Research on the ecological niches of LASV and better understanding of the diversity of mammarenaviruses offer more effective solutions for outreach efforts and prevention of Lassa fever.

References

Agbonlahor DE, Erah A, Agba IM, Oviasogie FE, Ehiaghe AF, Wankasi M, Eremwanarue OA, Ehiaghe IJ, Ogbu EC, Iyen RI, Abbey S (2017) Prevalence of Lassa virus among rodents trapped in three south-south states of Nigeria. J Vector Borne Dis 54(2):146–150. https://www.nimr.org.in/assets/542146.pdf%0A, https://ovidsp.ovid.com/ovidweb.cgi?T=JS&PAGE=reference&D=emexb&NEWS=N&AN=617660716

Akoua-Koffi C, Ter Meulen J, Legros D et al (2006) Detection of anti-Lassa antibodies in the Western Forest area of the Ivory Coast. Med Trop (Mars) 66(5):465–468

Andersen KG, Shapiro BJ, Matranga CB et al (2015) Clinical sequencing uncovers origins and evolution of Lassa virus. Cell 162(4):738–750. https://doi.org/10.1016/j.cell.2015.07.020

Arata AA, Gratz NG (1975) The structure of rodent faunas associated with arenaviral infections. Bull World Health Organ 52(4–5 6):621–627

Attinsounon CA, Ossibi Ibara BR, Alassani A et al (2018) Report of a fatal case of Lassa fever in Parakou in 2018: clinical, therapeutic and diagnostic aspects. BMC Infect Dis 18(1):1–4. https://doi.org/10.1186/s12879-018-3587-6

Bausch DG, Demby AH, Coulibaly M et al (2001) Lassa fever in Guinea: I. Epidemiology of human disease and clinical observations. Vector Borne Zoonotic Dis 1(4):269–281. https://doi.org/10.1089/153036601316977878

Blasdell KR, Duong V, Eloit M et al (2016) Evidence of human infection by a new mammarenavirus endemic to Southeastern Asia. Elife 5(JUN2016). https://doi.org/10.7554/eLife.13135

Bonner PC, Schmidt W-P, Belmain SR, Oshin B, Baglole D, Borchert M (2007) Poor housing quality increases risk of rodent infestation and Lassa fever in refugee camps of Sierra Leone. Am J Trop Med Hyg 77(1):169–175

Bonwitt J, Sáez AM, Lamin J et al (2017) At home with mastomys and rattus: Human-rodent interactions and potential for primary transmission of Lassa virus in domestic spaces. Am J Trop Med Hyg 96(4):935–943. https://doi.org/10.4269/ajtmh.16-0675

Borremans B, Leirs H, Gryseels S, Günther S, Makundi R, De Bellocq JG (2011) Presence of Mopeia virus, an African arenavirus, related to biotope and individual rodent host characteristics: Implications for virus transmission. Vector-Borne Zoonotic Dis 11(8):1125–1131. https://doi.org/10.1089/vbz.2010.0010

Borremans B, Vossen R, Becker-Ziaja B et al (2015) Shedding dynamics of Morogoro virus, an African arenavirus closely related to Lassa virus, in its natural reservoir host *Mastomys natalensis*. Sci Rep 5(1):10445. https://doi.org/10.1038/srep10445

Bowen MD, Peters CJ, Nichol ST (1997) Phylogenetic analysis of the arenaviridae: patterns of virus evolution and evidence for cospeciation between Arenaviruses and their rodent hosts. Mol Phylogenet Evol 8(3):301–316. https://doi.org/10.1006/mpev.1997.0436

Brown WM, George MJ, Wilson AC (1979) Rapid evolution of animal mitochondrial DNA. Proc Natl Acad Sci USA 76(4):1967–1971. https://doi.org/10.1073/pnas.76.4.1967

Carey DE, Kemp GE, White HA et al (1972) Lassa fever Epidemiological aspects of the 1970 epidemic, Jos, Nigeria. Trans R Soc Trop Med Hyg 66(3):402–408. https://doi.org/10.1016/0035-9203(72)90271-4

Central Intelligence Agency. The World Factbook. https://www.cia.gov/library/publications/resources/the-world-factbook/. Published 2019. Accessed 5 Oct 2019

Colangelo P, Verheyen E, Leirs H et al (2013) A mitochondrial phylogeographic scenario for the most widespread African rodent *Mastomys natalensis*. Biol J Linn Soc 108(4):901–916. https://doi.org/10.1111/bij.12013

Coulibaly-N'Golo D, Allali B, Kouassi SK et al (2011) Novel arenavirus sequences in hylomyscus sp. and Mus (Nannomys) setulosus from côte d'ivoire: Implications for evolution of arenaviruses in Africa. PLoS One 6(6). https://doi.org/10.1371/journal.pone.0020893

Demartini JC, Green DE, Monath TP (1975) Lassa virus infection in *Mastomys natalensis* in Sierra Leone. Gross and microscopic findings in infected and uninfected animals. Bull World Health Organ 52(4–5 6):651–663

Demby AH, Inapogui A, Kargbo K et al (2001) Lassa fever in Guinea: II. Distribution and prevalence of Lassa virus infection in small mammals. Vector Borne Zoonotic Dis 1(4):283–297. https://doi.org/10.1089/15303660160025912

Downs WG, Anderson CR, Spence L, Aitken THG, Greenhall AH (1963) Tacaribe virus, a new agent isolated from Artibeus bats and mosquitoes in Trinidad, West Indies. Am J Trop Med Hyg 12:640–646

Emmerich P, Günther S, Schmitz H (2008) Strain-specific antibody response to Lassa virus in the local population of west Africa. J Clin Virol 42(1):40–44. https://doi.org/10.1016/j.jcv.2007.11.019

Fichet-Calvet E, Lecompte E, Koivogui L et al (2007) Fluctuation of abundance and Lassa virus prevalence in *Mastomys natalensis* in Guinea West Africa. Vector-Borne Zoonotic Dis 7 (2):119–128. https://doi.org/10.1089/vbz.2006.0520

Fichet-Calvet E, Lecompte E, Koivogui L, Daffis S, Ter MJ (2008) Reproductive characteristics of *Mastomys natalensis* and Lassa virus prevalence in Guinea West Africa. Vector-Borne Zoonotic Dis 8(1):41–48. https://doi.org/10.1089/vbz.2007.0118

Fichet-Calvet E, Becker-Ziaja B, Koivogui L, Günther S (2014) Lassa serology in natural populations of rodents and horizontal transmission. Vector-Borne Zoonotic Dis 14(9):665–674. https://doi.org/10.1089/vbz.2013.1484

Forni D, Pontremoli C, Pozzoli U, Clerici M, Cagliani R, Sironi M (2018) Ancient evolution of mammarenaviruses: adaptation via changes in the L protein and no evidence for host-virus codivergence. Genome Biol Evol 10(3):863–874. https://doi.org/10.1093/gbe/evy050

Fraser DW, Campbell CC, Monath TP, Goff PA, Gregg MB (1974) Lassa fever in the Eastern province of Sierra Leone, 1970–1972. I. Epidemiologic studies. Am J Trop Med Hyg 23 (6):1131–1139. https://doi.org/10.4269/ajtmh.1974.23.1131

Gonzalez J-P, Duplantier JM (1999) The arenavirus and rodent co-evolution process: a global view of a theory. In: Saluzzo J-F, Dodet B, Foundation M-M (eds) Emergence and control of Rodent-Borne Viral diseases : Hantaviral and Arenal Diseases : Congrès, Annecy, 28–31 Octobre 1998. Elsevier, Paris

Gonzalez JP, Mccormick JB, Saluzzo JF et al (1983) An Arenavirus isolated from wild-caught rodents (Praomys Species) in the Central African republic. Intervirology 19(2):105–112. https://doi.org/10.1159/000149344

Gonzalez JP, Georges AJ, Kiley MP, Meunier DMY, Peters CJ, McCormick JB (1986) Evolutionary biology of a Lassa virus complex. Med Microbiol Immunol 175(2–3):157–159. https://doi.org/10.1007/BF02122439

Gryseels S, Rieger T, Oestereich L et al (2015) Gairo virus, a novel arenavirus of the widespread *Mastomys natalensis*: GENETICALLY divergent, but ecologically similar to Lassa and Morogoro viruses. Virology 476:249–256. https://doi.org/10.1016/j.virol.2014.12.011

Gryseels S, Baird SJE, Borremans B, Makundi R, Leirs H, Goüy de Bellocq J (2017) When viruses don't go viral: the importance of host Phylogeographic structure in the spatial spread of Arenaviruses. PLoS Pathog 13(1):1–22. https://doi.org/10.1371/journal.ppat.1006073

Haun BK, Kamara V, Dweh AS et al (2019) Serological evidence of Ebola virus exposure in dogs from affected communities in Liberia: a preliminary report. PLoS Negl Trop Dis 13(7):1–11. https://doi.org/10.1371/journal.pntd.0007614

ICTV (2018) Arenaviridae. International Committee on Taxonomy of Viruses. https://talk.ictvonline.org/ictv-reports/ictv_9th_report/negative-sense-rna-viruses-2011/w/negrna_viruses/203/arenaviridae. Published 2018. Accessed 16 May 2019

Keenlyside RA, McCormick JB, Webb PA, Smith E, Elliott L, Johnson KM (1983) Case-control study of *Mastomys natalensis* and humans in Lassa virus-infected households in Sierra Leone. Am J Trop Med Hyg 32(4):829–837. https://doi.org/10.4269/ajtmh.1983.32.829

Kenyon RH, McKee KTJ, Zack PM et al (1992) Aerosol infection of rhesus macaques with Junin virus. Intervirology 33(1):23–31. https://doi.org/10.1159/000150227

Khan SH, Goba A, Chu M et al (2008) New opportunities for field research on the pathogenesis and treatment of Lassa fever. Antiviral Res 78(1):103–115. https://doi.org/10.1016/j.antiviral.2007.11.003

Kouadio L, Nowak K, Akoua-Koffi C et al (2015) Lassa virus in multimammate rats, Côte d'Ivoire, 2013. Emerg Infect Dis 21(8):1481–1483. https://doi.org/10.3201/eid2108.150312

Kronmann KC, Nimo-Paintsil S, Guirguis F et al (2013) Two novel arenaviruses detected in pygmy mice Ghana. Emerg Infect Dis 19(11):1832–1835. https://doi.org/10.3201/eid1911.121491

Kunz S, Rojek JM, Roberts AJ, McGavern DB, Oldstone MBA, de la Torre JC (2006) Altered central nervous system gene expression caused by congenitally acquired persistent infection with lymphocytic Choriomeningitis virus. J Virol 80(18):9082–9092. https://doi.org/10.1128/jvi.00795-06

Lalis A, Wirth T (2018) Mice and men: an evolutionary history of Lassa fever. In: Biodiversity and evolution. Elsevier 189–212. https://doi.org/10.1016/b978-1-78548-277-9.50011-5

Lalis A, Leblois R, Lecompte E, Denys C, ter Meulen J, Wirth T (2013) The impact of human conflict on the genetics of *Mastomys natalensis* and Lassa virus in West Africa. PLoS One 7 (5). https://doi.org/10.1371/journal.pone.0037068

Lalis A, Evin A, Janier M, Koivogui L, Denys C (2015) Host evolution in *Mastomys natalensis* (Rodentia: Muridae): an integrative approach using geometric morphometrics and genetics. Integr Zool 10(6):505–514. https://doi.org/10.1111/1749-4877.12164

Lecompte É, Granjon L, Peterhans JK, Denys C (2002) Cytochrome b-based phylogeny of the Praomys group (Rodentia, Murinae): a new African radiation? Comptes Rendus—Biol 325 (7):827–840. https://doi.org/10.1016/S1631-0691(02)01488-9

Lecompte E, Fichet-Calvet E, Daffis S et al (2006) *Mastomys natalensis* and Lassa fever, West Africa. Emerg Infect Dis 12(12):1971–1974. https://doi.org/10.3201/eid1212.060812

Lecompte E, Aplin K, Denys C, Catzeflis F, Chades M, Chevret P (2008) Phylogeny and biogeography of African Murinae based on mitochondrial and nuclear gene sequences, with a new tribal classification of the subfamily. BMC Evol Biol. https://doi.org/10.1186/1471-2148-8-199

Leirs H (1994) Population ecology of *Mastomys natalensis* (Smith, 1834). Implications for rodent control in Africa

Leski TA, Stockelman MG, Moses LM et al (2015) Sequence variability and geographic distribution of Lassa Virus Sierra Leone. Emerg Infect Dis 21(4):609–618. https://doi.org/10.3201/eid2104.141469

Li K, Lin XD, Wang W et al (2015) Isolation and characterization of a novel arenavirus harbored by Rodents and Shrews in Zhejiang province China. Virology 476:37–42. https://doi.org/10.1016/j.virol.2014.11.026

Lukashevich IS, Clegg JC, Sidibe K (1993) Lassa virus activity in Guinea: distribution of human antiviral antibody defined using enzyme-linked immunosorbent assay with recombinant antigen. J Med Virol 40(3):210–217

Makundi RH, Massawe AW, Mulungu LS (2008) Reproduction and population dynamics of *Mastomys natalensis* Smith, 1834 in an agricultural landscape in the Western Usambara Mountains, Tanzania. Integr Zool 2(4):233–238. https://doi.org/10.1111/j.1749-4877.2007.00063.x

Malhotra S, Yen JY, Honko AN et al. (2013) Transcriptional profiling of the circulating immune response to Lassa virus in an aerosol model of exposure. PLoS Negl Trop Dis 7(4). https://doi.org/10.1371/journal.pntd.0002171

Manning JT, Forrester N, Paessler S (2015) Lassa virus isolates from Mali and the Ivory Coast represent an emerging fifth lineage. Front Microbiol 6:1037. https://doi.org/10.3389/fmicb.2015.01037

Mari Saez A, Cherif Haidara M, Camara A et al (2018) Rodent control to fight Lassa fever: evaluation and lessons learned from a 4-year study in Upper Guinea. PLoS Negl Trop Dis 12 (11):1–16. https://doi.org/10.1371/journal.pntd.0006829

Mariën J, Borremans B, Gryseels S et al (2017) Arenavirus dynamics in experimentally and naturally infected rodents. EcoHealth 14(3):463–473. https://doi.org/10.1007/s10393-017-1256-7

Mariën J, Kourouma F, Magassouba NF, Leirs H, Fichet-Calvet E (2018) Movement patterns of small rodents in Lassa fever-endemic villages in Guinea. EcoHealth 15(2):348–359. https://doi.org/10.1007/s10393-018-1331-8

Mariën J, Lo Iacono G, Rieger T, Magassouba N, Günther S, Fichet-Calvet E (2020) Households as hotspots of Lassa fever? Assessing the spatial distribution of Lassa virus-infected rodents in rural villages of Guinea. Emerg Microbes Infect 9(1):1055–1064. https://doi.org/10.1080/22221751.2020.1766381

Mccormick JB, Webb PA, Krebs JW et al (1987) A Prospective study of the epidemiology and ecology of Lassa Fever. J Infect Dis 155(3):437–444

Mills JN, Ellis BA, McKee KTJ et al (1992) A longitudinal study of Junin virus activity in the rodent reservoir of Argentine hemorrhagic fever. Am J Trop Med Hyg 47(6):749–763. https://doi.org/10.4269/ajtmh.1992.47.749

Monath TP (1975) Lassa fever: review of epidemiology and epizootiology. Bull World Health Organ 52(4–5 6):577–592

Monath TP, Newhouse VF, Kemp GE, Setzer HW, Cacciapuoti A (1974) Lassa virus isolation from *Mastomys natalensis* rodents during an epidemic in Sierra Leone. Science (80–) 185 (4147):263–265. https://doi.org/10.1126/science.185.4147.263

Mulungu LS, Ngowo V, Mdangi M et al (2013) Population dynamics and breeding patterns of multimammate mouse, *Mastomys natalensis* (Smith 1834), in irrigated rice fields in Eastern Tanzania. Pest Manag Sci 69(3):371–377. https://doi.org/10.1002/ps.3346

Nimo-Paintsil SC, Fichet-Calvet E, Borremans B et al (2019) Rodent-borne infections in rural Ghanaian farming communities. PLoS ONE 14(4):1–13. https://doi.org/10.1371/journal.pone.0215224

Ogunro B, Olugasa B, Kayode A, Kolawole S, Odigie EA, Happi C (2019) Detection of antibody and antigen specific For Lassa virus nucleoprotein in monkeys from Southern Nigeria: it is time to cast A Wider Zoo–surveillance Net. In: Nigerian centre for disease control. Abjua, Nigeria; 2019. https://lic.ncdc.gov.ng/downloads/o_a/O.A-4-Detection-Of-Antibody-And-Antigen-Specific-For-Lassa-Virus-Nucleoprotein-In-Monkeys-From-Southern-Nigeria-It-Is-Time-To-Cast-A-Wider-Zoo–surveillance-Net.pdf

Okoror LE, Esumeh FI, Agbonlahor DE, Umolu PI (2005) Lassa virus: seroepidemiological survey of rodents caught in Ekpoma and environs. Trop Doct 35(1):16–17. https://doi.org/10.1258/0049475053001912

Olayemi A, Obadare A, Oyeyiola A et al (2016a) Arenavirus diversity and phylogeography of *Mastomys natalensis* Rodents. Nigeria 22(4):13–16

Olayemi A, Cadar D, Magassouba N et al (2016b) New hosts of the Lassa virus. Sci Rep. 6 (1):25280. https://doi.org/10.1038/srep25280

Olayemi A, Oyeyiola A, Obadare A et al (2018) Widespread Arenavirus occurrence and seroprevalence in small mammals Nigeria. Parasites Vectors 11(1):1–9. https://doi.org/10.1186/s13071-018-2991-5

Olayiwola J, Bakarey A (2017) Epidemiological trends of Lassa fever outbreaks and insights for future control in Nigeria. Int J Trop Dis Heal 24:1–14. https://doi.org/10.9734/IJTDH/2017/32823

Oldstone MBA (2002) Biology and pathogenesis of lymphocytic choriomeningitis virus infection. Curr Top Microbiol Immunol 263:83

Parson W, Pegoraro K, Niederstätter H, Föger M, Steinlechner M (2000) Species identification by means of the cytochrome B gene. Int J Legal Med 114(1–2):23–28. https://doi.org/10.1007/s004140000134

Robbins CB, Krebs JW, Johnson KM (1983) Mastomys (rodentia: muridae) species distinguished by hemoglobin pattern differences. Am J Trop Med Hyg 32(3):624. https://doi.org/10.4269/ajtmh.1983.32.624

Roberts L (2018) Nigeria hit by unprecedented Lassa fever outbreak. Science 359(6381):1201–1202. https://doi.org/10.1126/science.359.6381.1201

Safronetz D, Sogoba N, Lopez JE et al (2013) Geographic distribution and genetic characterization of Lassa virus in Sub-Saharan Mali. PLoS Negl Trop Dis 7(12):4–12. https://doi.org/10.1371/journal.pntd.0002582

Saluzzo JF, Adam F, McCormick JB, Digoutte JP (1988) Lassa fever virus in senegal. J Infect Dis 157(3):605–605. https://doi.org/10.1093/infdis/157.3.605

Sayler KA, Barbet AF, Chamberlain C et al (2014) Isolation of Tacaribe Virus, a Caribbean Arenavirus, from host-seeking *Amblyomma americanum* ticks in Florida (Research Article). PLoS One 9(12). https://doi.org/10.1371/journal.pone.0115769

Schenk JJ, Rowe KC, Steppan SJ (2013) Ecological opportunity and incumbency in the diversification of repeated continental colonizations by muroid rodents. Syst Biol 62(6):837–864. https://doi.org/10.1093/sysbio/syt050

Shi M, Lin XD, Chen X et al (2018) The evolutionary history of vertebrate RNA viruses. Nature 556(7700):197–202. https://doi.org/10.1038/s41586-018-0012-7

Sogoba N, Feldmann H, Safronetz D (2012) Lassa fever in West Africa: evidence for an expanded region of Endemicity. Zoonoses Public Health 59(SUPPL.2):43–47. https://doi.org/10.1111/j.1863-2378.2012.01469.x

Stephenson EH, Larson EW, Dominik JW (1984) Effect of environmental factors on aerosol-induced Lassa virus infection. J Med Virol 14(4):295–303

Sun T, Vasek MJ, Klein RS (2014) Congenitally acquired persistent lymphocytic choriomeningitis viral infection reduces neuronal progenitor pools in the adult hippocampus and subventricular zone. PLoS ONE 9(5):1–8. https://doi.org/10.1371/journal.pone.0096442

Swanepoel R, Leman PA, Shepherd AJ et al (1985) Identification of Ippy as a Lassa-fever-related virus. Lancet 325(8429):639. https://doi.org/10.1016/S0140-6736(85)92175-0

Tagliapietra V, Rosà R, Hauffe HC et al (2009) Spatial and temporal dynamics of lymphocytic choriomeningitis virus in wild rodents, northern Italy. Emerg Infect Dis 15(7):1019–1025. https://doi.org/10.3201/eid1507.01524

Ter Meulen J, Lukashevich I, Sidibe K et al (1996) Hunting of peridomestic rodents and consumption of their meat as possible risk factors for rodent-to-human transmission of Lassa virus in the Republic of Guinea. Am J Trop Med Hyg 55(6):661–666

Theiler RN, Rasmussen SA, Treadwell TA et al (2009) Emerging and zoonotic infections in women. Infect Dis 377(2):364–377. https://doi.org/10.1016/j.idc.2008.05.007.Emerging

Vanden Broecke B, Borremans B, Mariën J et al (2018) Does exploratory behavior or activity in a wild mouse explain susceptibility to virus infection? Curr Zool 64(5):585–592. https://doi.org/10.1093/cz/zox053

Walker DH, Wulff H, Lange JV, Murphy FA (1975) Comparative pathology of Lassa virus infection in monkeys, guinea pigs, and *Mastomys natalensis*. Bull World Health Organ 52(4–5 6):523–534

Walker DH, McCormick JB, Johnson KM et al (1982) Pathologic and virologic study of fatal Lassa fever in man. Am J Pathol 107(3):349–356

Wilson DE, Reeder DM (1993) Mammal species of the world : a taxonomic and geographic reference, 2nd edn. Smithsonian Institution Press, Washington

Wulff H, Fabiyi A, Monath TP (1975) Recent isolations of Lassa virus from Nigerian rodents. Bull World Health Organ 52(4–5 6):609–613

Yadouleton A, Agolinou A, Kourouma F et al (2019) Lassa Virus in Pygmy Mice, Benin, West Africa, 2016–2017. Emerg Infect Dis 25(10):2016–2017. https://doi.org/10.3201/eid2510.180523

Epidemiology of Lassa Fever

Antoinette R. Bell-Kareem and Allison R. Smither

Contents

Abstract Lassa fever (LF) is a lethal hemorrhagic disease primarily concentrated in the tropical savannah regions of Nigeria and the Mano River Union countries of Sierra Leone, Liberia, and Guinea. Endemic hotspots within these countries have had recurrent exposure to Lassa virus (LASV) via continual spillover from the host

A. R. Bell-Kareem (✉) · A. R. Smither
Department of Microbiology and Immunology, School of Medicine, Tulane University, 1430 Tulane Avenue, JBJ 557, New Orleans, LA 70112, USA
e-mail: abell13@tulane.edu

Current Topics in Microbiology and Immunology (2023) 440: 87–109
https://doi.org/10.1007/82_2021_234

Published Online: 17 April 2021

reservoir *Mastomys natalensis*. Increased trade and travel throughout the region have spread the virus to previously unexposed countries, including Ghana, Benin, Mali, and Côte d'Ivoire. In the absence of effective treatment or vaccines to LASV, preventative measures against Lassa fever rely heavily on reducing or eliminating rodent exposure, increasing the knowledge base surrounding the virus and disease in communities, and diminishing the stigmas faced by Lassa fever survivors.

1 Introduction

Lassa virus (LASV), the causative agent of Lass fever, is an emerging virus that causes severe hemorrhagic fever. The word "emerging" may be a misnomer as evolutionary analysis unveiled the origin of LASV in Nigeria over 1000 years ago (Andersen et al. 2015). Additional clinical evidence of Lassa-like illness in the current endemic areas of West Africa has been documented as early as the 1930s (Monath 1975). Individuals contracted what was described as a savannah or murine typhus upon hunting or handling rodents and frequently succumbed to the disease (Monath 1975). A 1955–1956 outbreak of "encephalomyelitis of unknown origin" in Eastern Sierra Leone is described with strong similarities to Lassa fever with regards to sequalae, duration, and mortality (Rose 1957). As laboratory science improved, the ability to isolate and characterize novel infections allowed for the formal discovery of *Lassa mammarenavirus* and the diagnosis of Lassa fever in 1969 (Frame et al. (1970)). Reoccurring outbreaks, extensive domestic and international trade, and civil unrest has made LASV an emerging virus of upmost importance.

Lassa fever is endemic throughout much of West Africa. The predominant endemic areas of Lassa fever are Nigeria and the Mano River Union (MRU) countries of Liberia, Guinea, and Sierra Leone (Rose 1957; Monath et al. 1973). Extrapolating from seroprevalence studies in Sierra Leone conducted in the 1980s, it is estimated that 100,000–300,000 cases of Lassa fever occur annually resulting in 3,000–5,000 deaths (McCormick et al. 1987). The Kenema District of Sierra Leone has the highest incidence rate of Lassa fever in the world, with over 500 suspected cases presenting to the Kenema Government Hospital each year (Shaffer et al. 2014). Expanding diagnostic capabilities and mounting serological evidence point to the expansion of the Lassa fever zone to include bordering countries Burkina Faso, Mali, Benin, and Côte d'Ivoire as will be described in greater detail in later section (Fichet-Calvet and Rogers 2009). Herein, we describe the epidemiological characteristics of Lassa fever, epidemiology and LASV seroprevalence studies conducted throughout West Africa, and other societal factors influencing the stigma LF survivors face.

2 Epidemiology of Lassa Fever

2.1 Characteristics

Based on antibody seroprevalence data in the MRU and Nigeria, 80% LASV infections present asymptomatically or as a mild illness, mirroring those of other endemic diseases such as Typhoid fever or Malaria (Ogbu et al. 2007). Hospital care is not routinely sought out for minor febrile episodes making it difficult to assess accurate case counts. Of the potential hundreds of thousands of cases each year, only 20% of individuals will progress to severe disease (Ogbu et al. 2007). Within the 20% of fulminant cases, a 69% case fatality rate (CFR) is seen in most patients who present with viremia (Shaffer et al. 2014). However, CFR can range widely depending on the source of infection (i.e., nosocomial vs natural infection) (Kenmoe et al. 2020).

It is estimated that 20% of LASV infections arise from human-to-human transmission resulting from handling or encountering infected bodily fluids, a majority of which are from so-called "super-spreaders" (Lo Iacono et al. 2015). The remaining 80% of infections occur when individuals encounter urine, feces, and saliva shed from the rodent host *M. natalensis* (McCormick et al. 1987). This has been confirmed during the 2018 outbreak of Lassa fever in Nigeria, where viral genomes sequenced from patients showed a remarkable amount of diversity from one another, indicating continual spillover from the rodent population (Siddle et al. 2018).

Lassa fever appears to have a seasonal distribution, though cases can and do occur year-round. In West Africa, the rainy season consisting of short, frequent downpours generally spanning from May to November; the arid and hot dry season runs from December to April (Olayiwola and Bakarey 2017). There is a distinct uptick in cases at the beginning and end of the dry season, possibly due to the ability for increased group travel (Olayiwola and Bakarey 2017). As the rainy season progresses and roads are washed out or flooded, making it nearly impossible to travel long distances (Olayiwola and Bakarey 2017). Additionally, rodents face habitat and climate changes during the dry season that is more likely to drive them into homes, further increasing exposure risk (Mariën et al. 2018). Congregation of large groups of people alongside rodents increases the likelihood of density-dependent viral transmission.

2.2 Age/Sex Distribution

Overall, Lassa fever appears to equally target males and females, though particular outbreaks and countries may have a slight skew towards one sex or the other (Shaffer et al. 2014; Ilori et al. 2019; Monson et al. 1984). Studies within LASV endemic areas have reported varying age-specific attack rates and CFRs depending on the location and date of the outbreak and/or seroprevalence study (Shaffer et al. 2014; Abdulkarim et al. 2020; Bausch et al. 2001). Numerous reasons explain these differences: discrepancies in reporting, varying levels of surveillance, patients seeking care in hospitals versus their homes, and communities/health providers not properly attributing the ailment to Lassa fever (Shaffer et al. 2014; Bausch et al. 2001; Bagcchi 2020; Woyessa et al. 2019). However, research has shown that the presence of anti-LASV IgG increases with age, presumably as chances for exposure to the virus increases before antibody titers diminish (McCormick et al. 1987). Indeed, LASV antibodies have been detected decades after infection (Bond et al. 2013).

2.3 *Incidence in Pregnancy*

Contracting Lassa fever during pregnancy can lead to rapid demise of the mother, particularly during the third trimester where mortality rates are upwards of 80% (Price et al. 1988). Immediate evacuation of the uterus is paramount for increased chances of maternal recovery. The fetus rarely survives, although positive outcomes for both mother and baby are seen in sporadic instances (Agboeze et al. 2019; Branco et al. 2011; Monson et al. 1987). It was noted for the first time that pregnant women had a much higher mortality rate than the overall mortality rate for Lassa fever cases, 50% versus 17.9%–23.1% respectively, and a high likelihood of fetal loss upon recovery from Lassa fever (Keane and Gilles 1977). This was further confirmed in a 1987 study of pregnant and non-pregnant women with Lassa fever in a hospital located in Segbwema, Sierra Leone (Price et al. 1988). The study concluded a fetal and neonate loss of 87%, and that the CFR of non-pregnant women was lower than that of pregnant women, 13% versus 21% respectively (Price et al. 1988). Women in the third trimester held the highest maternal CFR with 30% (Price et al. 1988). This data along with other studies on maternal health during Lassa virus outbreaks demonstrates the extreme risk Lassa fever places on pregnant women, and potential death of the mother if the fetus is not evacuated in time (Agboeze et al. 2019; Branco et al. 2011; Monson et al. 1987).

3 Incidence in the Lassa Fever Zone

3.1 *Nigeria*

The very first confirmed cases of Lassa fever originated in Nigeria. In 1969, an American nurse working at the Lassa Mission Hospital, came down with an enigmatic illness (Frame et al. 1970). She was sent to Bingham Hospital in Jos for further treatment. In Jos, she infected two missionary nurses caring for her.

The nurse from Lassa and one of the nurses in Jos died from the disease. The third nurse, Lily Pinneo, was flown to New York Presbyterian Hospital for further evaluation and treatment (Watts 2012). Aliquots of her blood were sent to Yale University, where investigator Jordi Caslas-Ariet and his lab isolated and characterized the virus in her blood (Buckley and Casals 1970). Dr. Casals-Ariet and a lab technician from another lab in the facility fell ill, and though Dr. Casals-Ariet recovered, the lab technician died from the illness (Saltzmann 1978). Upon electron microscopy of the virus, it was determined that it resembled the related, rodent-borne *Arenaviridae* member Lymphocytic Choriomeningitis virus (LCMV) yet had low complement-fixing reactivity to it (Buckley and Casals 1970). The hitherto undescribed virus was named after the tiny village in northeast Nigeria where it originated (Saltzmann 1978). In the years following the initial outbreak, another outbreak of Lassa fever was seen in Jos and surrounding areas of the former Benue-Plateau State, now known as Plateau State (Carey et al. 1972). During this 1970 outbreak, the 28 suspected cases were thought to be linked by contact with the index cases, and 13 of the suspected individuals died (Carey et al. 1972). Additional cases appeared in Onitsha, Anambra State, in 1974, expanding the range of Lassa fever to southern Nigeria (Bowen et al. 1975).

Documented outbreaks extending from the mid 1970′s to the early 1990′s has shown the expansion of LASV from towns surrounding Jos to many urban and rural areas of southern and northern Nigeria (Fig. 1). In the early 2000′s, sporadic outbreaks were concentrated mostly in the Edo and Ondo states (Omilabu et al. 2005; Ehichioya et al. 2010). Since then, a gradual uptick in the annual incidence of Lassa fever increased from a few dozen cases a year in one or two states to several hundred confirmed cases throughout Nigeria (World Health Organization 2021; Nigeria Centre for Disease Control 2020a). In response, the Nigerian Center of Disease Control activated emergency operations to the 2018 and 2019 Lassa Fever outbreaks (Nigeria Centre for Disease Control 2018, 2019). Nearly 2000 suspected cases and 246 confirmed cases were reported in 2017, a 60% increase from 2016 (Akpede et al. 2019). By the end of 2020, the Nigerian Center of Disease Control reported 1181 confirmed cases along with 244 confirmed deaths, a CFR of 20.7% (Nigeria Centre for Disease Control 2020b). This year-end report illustrated that Lassa fever outbreaks were recorded from 27 States, and this distribution reached every geopolitical zone of Nigeria (Nigeria Centre for Disease Control 2020b). The rise in cases over the past decade likely corresponds with improved diagnostic capacity within Nigeria, rather than a genuine increase in the number of infections occurring.

Although much of the molecular epidemiology in Nigeria has focused on the phylogeography and phylogenetics of the circulating LASV strains, passive surveillance of Lassa fever between outbreaks has aided in characterizing the seroprevalence of LASV in Nigeria. The first seroprevalence study undertaken in 1987 found an overall seroprevalence of 21% among the tested states for LASV with prevalence rates from 13% in the former Gongola State, now Adamawa and Taraba States, to 38% in Benue State (Tomori et al. 1988). A 1997 seroprevalence survey of 552 healthcare workers in six different facilities in the Plateau, Nasarawa,

Fig. 1 **Chronological distribution of Lassa fever cases in Nigeria**: (Map marker, location, date of outbreak/case) 1. Rahama 1952; 2. Lassa 1969; 3. Bassa 1970; 4. Jos 1970, 1972, 1973, 1992–93, 2016; 5. Vom 1970, 1974 -75, 1976, 1977; 6. Sokoto 1971; 7. Takum 1974; 8. Onitsha Hospital 1974; 9. Zonkwa 1975; 10. Fan 1977; 11. Gboko 1987; 12. Ondo 1987; 13. Gongola 1987; 14. Lafia Hospital 1987, 1992–93; 15. Aba Hospital 1989; 16. Aboh Mbaise Hospital 1989; 17. Maiduguri 1992–93; 18. Ekpoma 2001–04; 19. Ibilo 2001–04; 20. Igarra 2001–04; 21. Jalingo 2007; 22. Irrua 2009–2010,2009–2018, 2011–2015; 23. Abakaliki 2012, 2018; 24. Taraba State 2015–16; 25. Bauchi 2016–18; 26. Edo & Ondo States 2018–19

and Borno states found seroprevalence was highest in ward aids whose jobs warranted the most contact with infected patients, with seroprevalence ranging from 1.3% (Borno) to 27.3% (Nasarawa) (Bajani et al. 1997). Similar seroprevalence among healthcare workers and the general populace have been observed in subsequent studies (Olayiwola and Bakarey 2017; Omilabu et al. 2005; Emmerich et al. 2008; Bukbuk et al. 2014; Richmond and Baglole 2003).

3.2 Sierra Leone

The Eastern Province of Sierra Leone is considered to have the highest concentration of endemic Lassa fever in the world (Shaffer et al. 2014). As mentioned previously, Lassa-like illnesses have been described in the area since the early 1930's, though the first official cases of Lassa fever in Sierra Leone were diagnosed in 1972 (Monath 1975). Unlike the concurrent outbreaks in Nigeria and Liberia, most Lassa fever cases in Sierra Leone stemmed from non-nosocomial sources (Ogbu et al. 2007; Keane and Gilles 1977). The 1970–1972 outbreak in Panguma and Tongo saw 63 suspected cases with 24 deaths, an overall CFR of 38.1% (Ogbu et al. 2007). Some contacts of patients during the 1972 outbreak developed antibodies to LASV, indicating for the first time that not all LASV infections result in Lassa fever (Fabiyi and Tomori 1975). The next documented outbreak in Sierra Leone occurred in hospitals located in Panguma and Segbwema between 1973 and 1976 (Keane and Gilles 1977). The Panguma hospital confirmed 40 patients from January to March 1973, and a CFR of 17.9%, while the Segbwema hospital confirmed 11 patients in 1975 and had a CFR of 75% (Keane and Gilles 1977). Together the outbreaks formed a combined CFR of 50% (Keane and Gilles 1977). Another research group in the 1980's, performed the first seroprevalence studies in Sierra Leone, finding 26% of the population tested in the Eastern Province having antibodies to LASV, with higher prevalence in mining towns and villages where interactions with rodents may be more frequent (Keenlyside et al. 1983).

The Sierra Leonean Civil War from 1991–2003 decimated the emerging research infrastructure, displaced local healthcare workers and scientists, and severely limited accurate record keeping. However, a World Health Organization report indicated 823 cases of Lassa fever from January 1996-April 1997 in the Kenema District, with 153 deaths (CFR 18.6%) (Richmond and Baglole 2003). Members of the Sierra Leone army and UN peacekeepers contracted Lassa fever during the outbreak (ter Meulen et al. 2001). The high incidence of cases can be attributed to the destruction of homes and lifestyles due to the war, which may have led to war survivors displaced to forested areas for shelter, larger groups of people living in closer proximity, and resorting to alternate diets that included rodents when other food sources were scarce (Lassa fever Resources 2020). Lassa fever research resumed in 2004 with the formation of the Mano River Union Lassa Fever Network (MRU-LFN), an international collaboration between Tulane University, the WHO, US Army Medical Research Institute of Infectious Disease, and other non-government organizations based out of Kenema Government Hospital (KGH) (Shaffer et al. 2014; Khan et al. 2008). In the years following the civil war and prior to the development of diagnostic assays (2004–2008), over 400 probable or confirmed cases were admitted to the Lassa fever ward at KGH (Shaffer et al. 2014). Since the deployment of diagnostic antigen immunoassays in 2008, the number of admitted patients to the ward has decreased (Branco 2011). However, Lassa fever remains prevalent through Sierra Leone, with 10/13 districts having at least one confirmed case between 2008 and 2012 (Shaffer et al. 2014). Sierra Leone continues to report ongoing cases of Lassa fever in endemic areas. A recent outbreak in 2019 resulted in a fatal exported case in the Netherlands (Disease outbreak news 2019; Overbosch et al. 2020). Documented cases and outbreaks in Sierra Leone are depicted in Fig. 2.

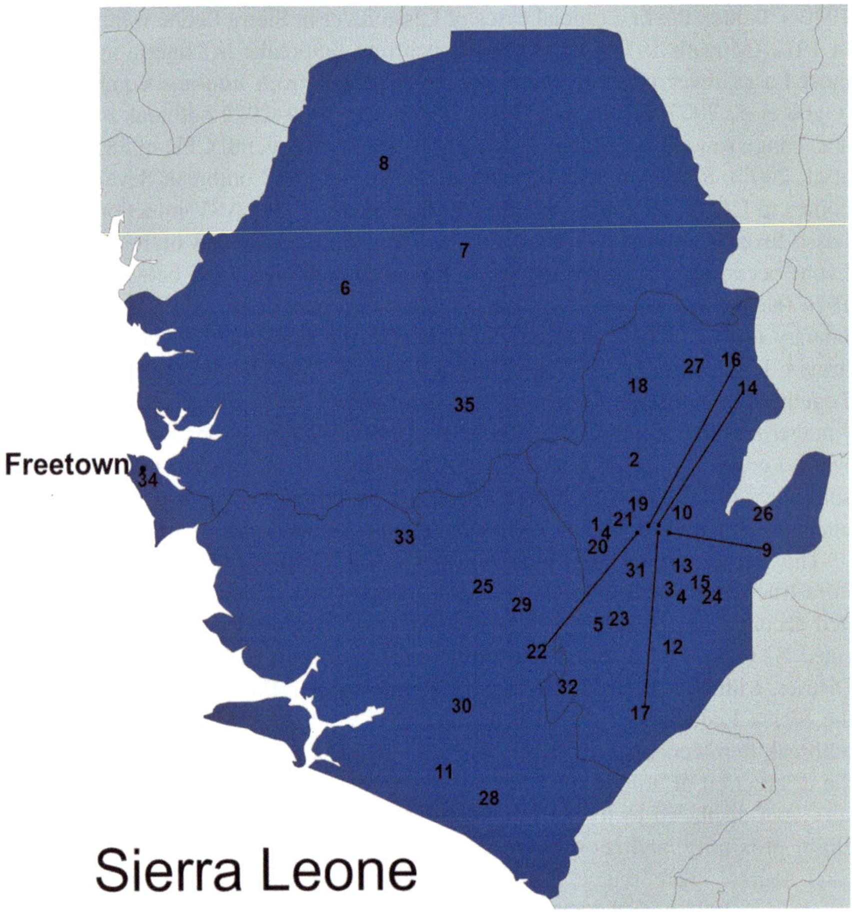

Fig. 2 Chronological Distribution of Lassa fever Cases in Sierra Leone: (Map marker, location, date of outbreak/case) 1. Panguma Hospital 1970–75, 1996–97, 2003; 2. Tongo Hospital 1970–72, 1996–97; 3. Segbwema Hospital 1975, 1981–85, 1985, 1996–97; 4. Panguma & Segbwema 1977–81; 5. Serabu Hospital 1977; 6. Mamaka 1977–82; 7. Kamabunyele 1977–83; 8. Kathumpe 1977–83; 9. Lowoma 1977–82; 10. Konia 1977–83; 11. Kpandebu 1977–83; 12. Neama 1977–83; 13. Niahun 1977–83; 14. Njakundoma 1977–83; 15. Semewabu 1977–83; 16. Bomie/Landoma 1977–83, 1996–97; 17. Palima/Tongola 1977–83, 1996–97; 18. Buima 1996–97; 19. Daabu 1996–97; 20. Macca 1996–97; 21. Pandebu 1996–97; 22. Tokpombu 1996–97; 23. Kenema Hospital 1996–97, 1999, 2001–04, 2008–2016, 2018, 2019; 24. Daru Hospital 2000; 25. Bo 2001, 2011–12, 2018; 26. Kailahun Hospital 2001; 27. Kono Hospital 2001; 28. Pujehun Hospital 2001; 29. Gerihun Camp 2003; 30. Jimmi Camp 2003; 31. Largo Camp 2003; 32. Nongowa 2003; 33. Taiama Camp 2003; 34. Freetown 2014–15; 35. Masanga 2018

3.3 Liberia

Liberia is a heavily forested country bordering Guinea to the south, Sierra Leone to the east, and Côte d'Ivoire to the west. Lassa fever is endemic in Liberia, with most cases converging in the north-central and northwest regions of the country. These regions share similar geographic features to its neighbors Guinea and Sierra Leone (Olugasa and Dogba 2015). Lassa fever has also been reported in the northern portion of coastal Grand-Bassa county (Olugasa and Dogba 2015). The first recorded outbreak of Lassa Fever was reported in 1972 in Zorzor during a simultaneous outbreak in Sierra Leone from 1970 to 1972 (Fichet-Calvet and Rogers 2009; Fraser et al. 1974). The index case of the Zorzor outbreak was admitted to the general obstetrics ward of the hospital (Monath et al. 1973). From there, the virus spread to three pregnant patients and seven caregivers within a week of her discharge from the hospital (Monath et al. 1973). 4 of the 10 additional patients died (Monath et al. 1973). This outbreak pointed to how greater exposure to the virus within a hospital by caregivers/nurses increased their chances of becoming infected (Monath et al. 1973). Many outbreaks have been reported since the first outbreak in Zorzor, with the most recent report in Suakoko in 2019 (World Health Organization Regional Office for Africa 2019) (Fig. 3).

Most clinical and seroprevalence studies of Lassa fever in Liberia were conducted in the late 1970′s to the mid 1980′s (Monson et al. 1984; Frame et al. 1984). Prevalence of LASV seems to be lower among the general population than in other countries, ranging from 1.9% in deeply forested villages to 6.4% in roadside villages of the northwest portion of Liberia (Yalley-Ogunro et al. 1984). Presence of anti-Lassa IgG antibodies in healthcare workers, who likely have much more exposure to potentially infected individuals, varied in different regions of the country ranging from 3.8% in the southeast coastal region to 40.4% in Lofa County bordering Sierra Leone and Guinea (Frame et al. 1984). The Liberian civil wars spanning from 1989 to 2003 decimated the biomedical research infrastructure and prevented more up-to-date surveys from being conducted. As peace continues in the region, newer seroprevalence studies with updated technology in combination with advanced record keeping at local and national levels are needed to reassess the true prevalence of Lassa fever in the country (Olugasa and Dogba 2015; Kruk et al. 2010).

3.4 Guinea

Guinea is the largest of the three MRU countries. It borders Sierra Leone and Liberia to the south, Côte d'Ivoire to the east, Mali to the north, and Senegal and Guinea-Bissau to the west. Much of the forested land lies in the southeastern part of the country near the borders of Sierra Leone and Liberia. This contiguous tropical forest stretching between the three countries is an endemic hot zone of Lassa fever.

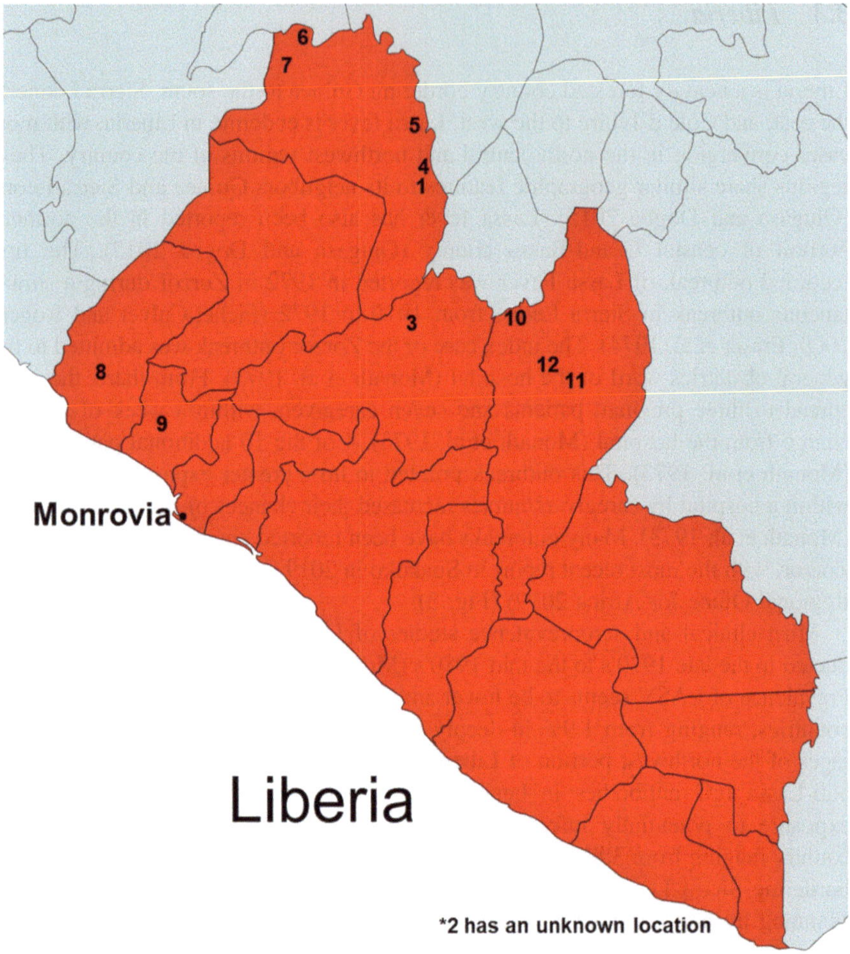

Fig. 3 Chronological Distribution of Lassa fever Cases in Liberia: (Map marker, location, date of outbreak/case) 1. Zorzor Hospital 1969, 1972, 1977, 1979, 1980–1982; 2. Hayes Gowien Town Mission 1971; 3. Suakoko 1972, 2016, 2019; 4. Yielah 1972; 5. Zigida 1972; 6. Koindu 1974; 7. Foya Kamara Hospital 1977, 1979, 1980, 1981; 8. Mano River Hospital 1980; 9. Goodrich Plantation Hospital 1980; 10. Ganta Hospital 1982, 2004, 2018; 11. Louplay 2006; 12. Saglelpie 2007

Indeed, a 2001 study identified that over 40% of recently febrile patients is hospitals in this region were IgG positive for LASV (Klempa et al. 2013). Compared to the rolling hills of forest in the southeast, Lassa fever is rarely seen in the flat coastal plains in the southern coast, savannah, or the mountainous central part of the country (Woyessa et al. 2019). The geographic diversity of Guinea along with its relative political stability has allowed for extensive epidemiology and ecology studies to be conducted.

Lassa fever in Guinea has been documented since the early 1980′s, though an outbreak of Lassa fever has never been declared in the country (Knobloch et al. 1980; Boiro et al. 1987) (Fig. 4). The first extensive human seroprevalence studies were conducted in mid-1990′s (Lukashevich et al. 1993). Human prevalence of LASV IgG was highest in tropical secondary forests and savannahs (25%-55%), and lowest in the mountainous and coastal areas (4–7%) (Lukashevich et al. 1993). There were no differences in prevalence between males and females, nor amongst age groups (Lukashevich et al. 1993). Interestingly, there is great village-to-village, and potentially year-to-year variability even within areas of high LASV prevalence (Lukashevich et al. 1993; Meulen et al. 1996). Over a decade later, a study was conducted in the southern tropical forest areas and it reported overall seroprevalnece rates at 11.3%, despite sampling from some of the same villages (Kernéis et al. 2009). The most recent seroprevalence study conducted in the Faranah prefecture of southern Guinea show very high seroprevalence with some villages having almost their entire population IgG positive for LASV (Magassouba et al. 2018). Differences in methods of detection may explain the wide varibility observed over the years. As diagnostic capacity has increased and assays become more sensitive, it is possible that earlier studies masked or under detected true positives. Likewise, there is still a chance of false positives even with the most sophosticated diagnostic test.

4 Expansion of the Lassa Fever Zone: Other West African Countries

4.1 Precedent

Beyond the aforenoted countries, there have been suspected cases, confirmed diagnoses, and serological evidence of Lassa fever in other West African countries. A 1974 surveillance study found antibodies against LASV in missionaries stationed outside the traditional endemic zone, including Burkina Faso, the Central African Republic, the Democratic Republic of Congo, Côte d'Ivoire, and Mali (Frame 1975) (Fig. 5). These countries all have areas of tropical savannah like that of other countries in the endemic zone, an environment conducive to the persistence of rodents harboring LASV (Sogoba et al. 2012). The spread from Lassa, Nigeria westward to other countries is likely due the contiguity of niche habitation and subsequent movement of the *Mastomys* rodent (Richmond and Baglole 2003; Meulen et al. 1996). When comparing four of the reported lineages of LASV to an emerging fifth lineage in Mali and Côte d'Ivoire, a group suggested that usual trade between culturally similar countries and the colonial movement of enslaved people from the middle of Africa to the west coast attributed the presence of virus to these countries (Manning et al. 2015). Some credit the diaspora of refugees from Sierra Leone during its civil war from 1992 to 2002, which not only blocked medical

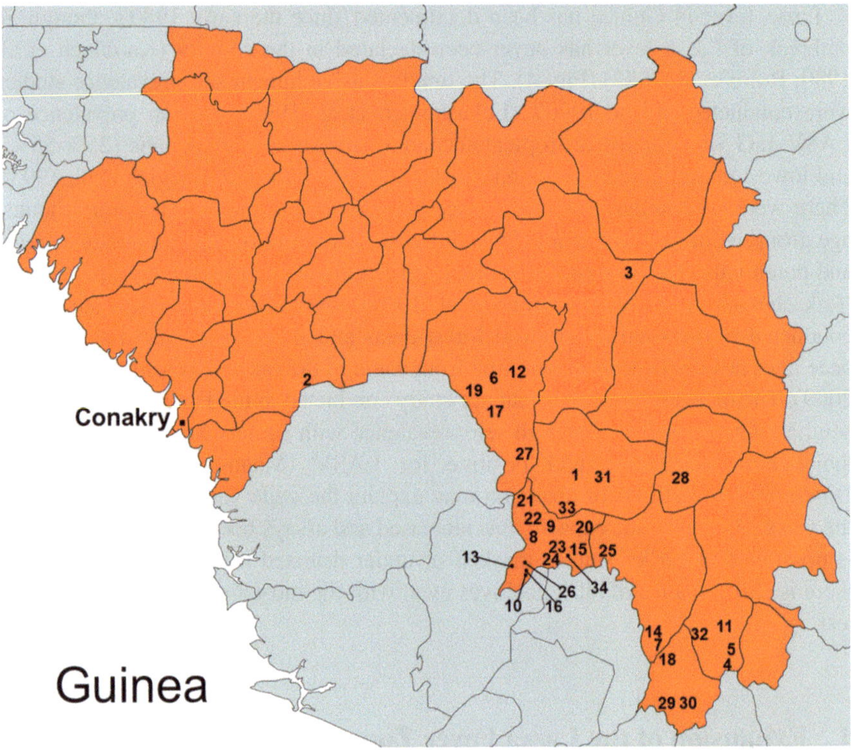

Fig. 4 Chronological Distribution of Lassa fever Cases in Guinea: (Map marker, location, date of outbreak/case) 1. Teleokoro 1965, 1967, 1968, 1996–99; 2. Madina Oula 1982–83, 1990–92, 1996–97; 3. Balato 1990–92; 4. Gbah 1990–92; 5. Gbenemou 1990–92; 6. Kamaraya 1990–92; 7. Komore 1990–92; 8. Nongoa Mbalia 1990–92; 9. Sassani Toli 1990–92; 10. Telekolo 1990–92; 11. Thuo 1990–92; 12. Tindo 1990–92; 13. Fangamandou 1990–92, 1993, 1996–97; 14. Bamakama 1990–92, 1996–97; 15. Bawa 1990–92, 1996–97; 16. Owe Jiba 1990–92, 1996–97; 17. Sangoyah 1990–92, 1996–97; 18. Waita 1990–92, 1996–97; 19. Gbetaya 1990–92, 1996–97, 2003–05; 20. Guedembou 1993; 21. Kassadou 1993; 22. Kpolodou 1993; 23. Temessadou 1993; 24. Tomandou 1993; 25. Tekoulo 1993, 1996–97; 26. Denguedou 1993, 2005; 27. Bambaya 1996–99; 28. Banankoro 1996–99; 29. Bignamou 1996–99; 30. Dieke 1996–99; 31. Fedou 1996–99; 32. Koulenin 1996–99; 33. Yende Milimou 1996–99; 34. Guéckédou Hospital 2020

resources and research, but also pushed its citizens to non-endemic countries like Côte d'Ivoire, Burkina Faso, and Mali that later documented cases of Lassa Fever (Fichet-Calvet and Rogers 2009; Richmond and Baglole 2003; Manning et al. 2015; Schmitz et al. 2002).

The first reported cases of Lassa fever outside the Lassa fever zone were in the Benin in 1977, and Upper Volta (now Burkina Faso) in 1980 (Saltzmann 1978; World Health Organization 1981). The case from Benin was from a lab technician at a hospital in Bembèrèkè who may have been exposed to suspected LASV cases a few months prior, and ultimately succumbed to the disease (Saltzmann 1978). The

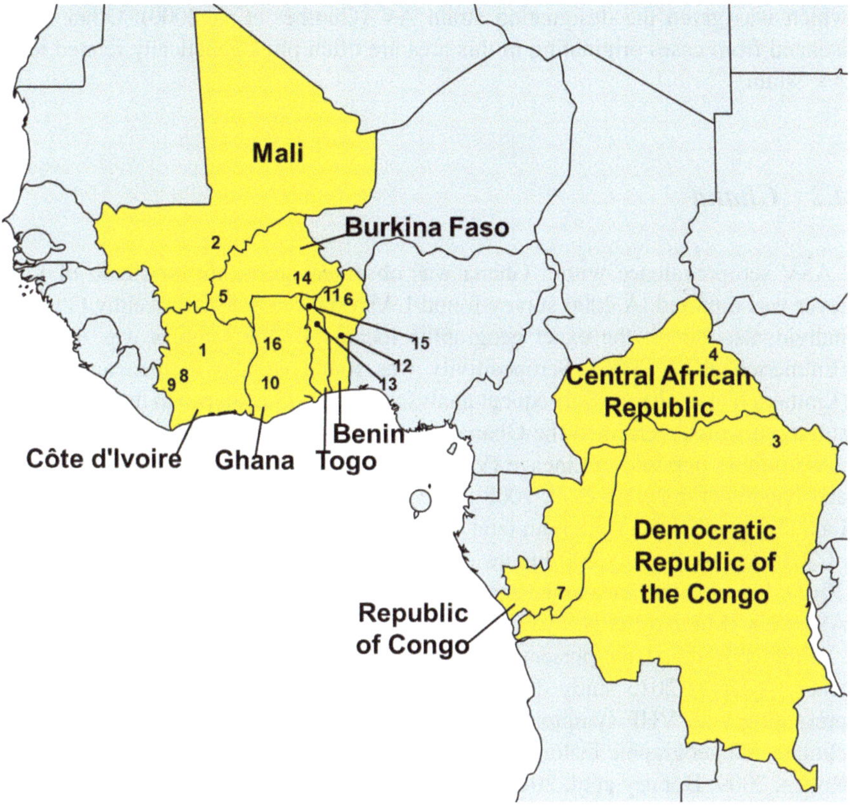

Fig. 5 Chronological Distribution of Lassa fever Cases in Non-Endemic African Countries: (Map marker, location, date of outbreak/case) 1. Beoumi, Côte d'Ivoire 1970–74; 2. Ntorosso, Mali 1971; 3. Banda, Zaire (Democratic Republic of the Congo) 1972; 4. Djema, Central African Republic 1972; 5. Banfora, Burkina Faso 1974; 6. Bambéréké Hospital, Benin 1977; 7. Ngamambou, Republic of the Congo 1981; 8. Forêt Classée Côte d'Ivoire 2000; 9. Guiglo, Côte d'Ivoire 2000; 10. Ghana 2011; 11. Tanguieta, Benin 2014; 12. Tchaourou, Benin 2016, 2017; 13. Mango, Togo 2016; 14. Ouargaye, Burkina Faso 2017; 15. Kpendjal, Togo 2017; 16. Kintampo, Ghana 2018

second case stemmed from a worker who had previously traveled to Burkina Faso and was later admitted to a hospital in Amsterdam with a fever of unknown origin (World Health Organization 1981). The patient recovered, but high titers of anti-Lassa antibody were detected post-discharge (World Health Organization 1981). A third case was reported when a German student travelling between Côte d'Ivoire, Ghana, and Burkina Faso in 2000 contracted the virus (Günther et al. 2000). She was eventually diagnosed with Lassa fever and passed despite immediate administration of ribavirin and supportive care (Günther et al. 2000). The strain of virus from this patient was determined to be a separate lineage of LASV that had distinct characteristics from strains circulating in the MRU and Nigeria, which was given the designation strain AV (Günther et al. 2000). Other strains isolated from cases originating in this area are often phylogenetically related to the AV strain.

4.2 Ghana

LASV seroprevalence within Ghana was observed before the first case of Lassa fever was reported. A 2006 survey found LASV IgG in 3–4% of healthy Ghanaian individuals, though the exact geographic locality of the samples was not listed (Emmerich et al. 2006). Seropositivity was found equally in men and women (Emmerich et al. 2006). Subsequent analyses of the IgG response to different LASV strain antigens determined the Ghanaian samples to be most closely related to the AV strain, as opposed to Lineage IV circulating in the MRU or Lineage III in Benin and Nigeria (Emmerich et al. 2008). The initial two laboratory confirmed cases of Lassa fever in Ghana were both fatal and originated in the Ashanti region at the end of 2011, approximately 200 km away from the Côte d'Ivoiran border (Dzotsi et al. 2012). It was later determined that both were phylogenetically clustered with the AV strain (Manning et al. 2015).

Further studies have presented conflicting results with respect to LASV prevalence. A larger 2013 study did not find evidence of LASV infection in patients presenting with VHF symptoms in north and central Ghana, areas with presumed climate and geographic features hospitable to LASV persistence (Fichet-Calvet and Rogers 2009; Bonney et al. 2013). In 2019, adult human serum from other villages in the same general regions indicated overall LASV seroprevalence to be 5.2%, with individual villages as high as 16% (Nimo-Paintsil et al. 2019).

4.3 Benin

Despite, Benin's first Lassa fever case in 1977, LASV research in the country did not begin until the mid-2000′s (Saltzmann 1978). Current serological evidence points to LASV circulating with seroprevalence rates around 10% in 2008 (Emmerich et al. 2008). The first laboratory confirmed case of Lassa fever occurred in the northwest corner of the country in November 2014 during the West African Ebola outbreak (World Health Organization 2015). An outbreak of Lassa fever during the first half of 2016 was concentrated in the northeast district of Borgou and saw 54 suspected and 16 confirmed cases, with 28 deaths (CFR of 51.8%) (World Health Organization 2015, 2016a, b). Analysis of blood samples collected from these cases showed the viral strains were related to Lineage I and II in Nigeria (Salu et al. 2019). Lassa fever was again reported in the Borgou district in a pregnant woman in February of 2017 (World Health Organization 2017). The woman delivered via cesarean section prior to passing (World Health Organization 2017).

The neonate was taken back to northern Togo by the father, where the baby tested positive for LASV and was successfully treated with ribavirin (World Health Organization 2017). In the same year, a Benin native working in Nigeria became ill and decided to return to his familial home in Togo (Centers for Disease Control and Prevention 2019). On his way he traveled through Benin, stopping at a hospital before reaching his family in Nadoba, Togo to see healers (Centers for Disease Control and Prevention 2019). He then returned to Benin to seek help from another healer, and within a month of his symptoms beginning he died in a Benin hospital (Centers for Disease Control and Prevention 2019). His story highlights how individuals can unknowingly carry Lassa fever, thus increasing the chance of outbreaks in multiple areas.

4.4 Mali

Evidence of Lassa fever in southern Mali has been seen since the late 2000′s. A fatal case phylogenetically related to the AV strain from Ghana, Burkina Faso, and Côte d'Ivoire, was exported to the United Kingdom from the village of Soromba bordering northern Guinea in 2009 (Safronetz et al. 2010). While investigating this case, several *M. natalensis* rodents captured in Soromba and two other nearby villages were found to have a strain of the virus virtually identical to the 2009 patient's strain (Safronetz et al. 2010). Follow-up visits to these villages indicate presence of LASV IgG to be between 14.5–44%, with seroconversion seen in 6.3% of individuals (Sogoba et al. 2016; Safronetz et al. 2017). LASV IgG was found equally in both males and females, with seropositivity increasing with age (Sogoba et al. 2016; Safronetz et al. 2017).

In 2019, two cases of Lassa fever were reported in the capital of Bamako, one genetically clustered with Lineage IV found in Guinea, and the other with the Soromba strain from 2009, potentially indicating circulation of multiple strains of LASV within Mali (Baumann et al. 2019). Phylogenetic analysis of the Soromba strain groups this isolate with strain AV and other Malian, Ghanian, and Côte d'Ivorian strains of LASV (Manning et al. 2015). This divergence from Lineage IV occurred 250–300 years ago, potentially indicating a distinct lineage of LASV in these areas (Manning et al. 2015).

4.5 Other Countries

Côte d'Ivoire has not experienced a confirmed case of Lassa fever during the prodromal phase, despite LASV circulating in the rodent population (Mateo et al. 2019). Serological evidence of LASV has been demonstrated particularly in the western half of the country, with seroprevalence rates among loggers in the area over 25% (Emmerich et al. 2008; Akoua-Koffi et al. 2006). Interestingly, a retrospective analysis of Yellow fever negative febrile patients identified a fatal case of Lassa fever in 2015 that grouped with Lineage IV strains found in Sierra Leone and Guinea (Mateo et al. 2019). Since the travel history of the patient is unknown, it is undetermined whether the case was locally acquired or imported (Mateo et al. 2019).

Togo has only reported three occurrences of Lassa fever, one of which was imported from Benin as described above (World Health Organization 2017; Centers for Disease Control and Prevention 2019). One clustered outbreak was reported in 2016 (Patassi 2017). A German surgeon at a private hospital in northern Togo referred himself to the same hospital with fever (Patassi et al. 2017). The fever failed to resolve upon administration of anti-malarial treatment, and the patient gradually regressed (Patassi et al. 2017). An American nurse caring for the patient also became sick (Patassi et al. 2017; McElroy et al. 2017). The patient was airlifted to Germany, and the nurse to the United States (McElroy et al. 2017). The nurse survived after administration of ribavirin, but the patient died. An undertaker in Germany then acquired the virus from the patient post-mortem (Lehmann et al. 2016). This was the first case of locally transmitted LASV outside of West Africa. The presumed source of the infection to the index case was never confirmed but may have been acquired from a Togolese patient the surgeon operated on (Patassi et al. 2017). Sequencing analysis of isolated virus from this case cluster demonstrated resemblance somewhere between Lineage I (Pinneo) and newly discovered rodent strains in Nigeria (Whitmer et al. 2018). This Togo strain is being potentially proposed as a new lineage of LASV (Whitmer et al. 2018). Two other cases of exported Lassa fever were reported in 2017 (World Health Organization 2017).

Burkina Faso has only reported two historic cases of Lassa fever, as previously mentioned above, until 2017 (World Health Organization 2017). One of the exported cases of Lassa fever seen in Togo was originally from the Ouargaye district of Burkina Faso (World Health Organization 2017). The patient was a pregnant woman who was initially hospitalized twice in Burkina Faso, and after the second hospitalization she was transferred to a northern Togo hospital where she expired (World Health Organization 2017). No epidemiology or ecology studies have been conducted in Burkina Faso to date.

Senegal has had no confirmed cases of Lassa fever, however, there is serological evidence from the early 1970′s that infection was present in the eastern part of the country (Monath 1975). A 1988 study on the seroprevalence of LASV in rodents and people from various parts of the country revealed that none of the human sera contained antibodies against LASV and only 18 out of 1440 rodent samples were positive for antibodies, mostly from the *Mastomys* species (Whitmer et al. 2018).

The Central African Republic, CAR, has never reported a case of Lassa fever; however, the nation has been involved in serological studies during the 1980′s that pointed to possible LASV infections and/or seroprevalence (Gonzalez et al. 1989). The results of these studies have since been called into question upon the discovery of Ippy virus, another rodent-borne mammarenavirus found in CAR (Swanepoel et al. 1985). The cross-reactivity between Ippy virus and LASV may have attributed to the results of these earlier studies when diagnostic tools were not as stringent to detect cross-reactivity among closely related viruses (Meunier et al. 1985).

4.6 Exported Cases

There have been 33 cases of Lassa fever exported outside of West Africa as of 2019, making LASV the most prevalent category A bioterrorism agent globally (Kofman et al. 2019). Passengers unknowingly carrying Lassa fever may have inadvertently spread the virus to hundreds of travelers, which could have profound public health consequences. Only one case of asymptomatic secondary transmission has arisen from exported cases outside of West Africa (Kofman et al. 2019). Despite this, it is of upmost importance for port authorities and airlines to assess and surveille passengers in transit for febrile symptoms. Below, we highlight key cases of Lassa fever exported to other countries not described in previous sections.

There have been eight imported cases of Lassa Fever to the United States, three of which were patients medically evacuated from West Africa. Lily Pinneo, the missionary nurse stationed in Lassa, Nigeria, was medically evacuated for treatment in the US in 1969 (Saltzmann 1978). In 1976, a Peace Corps member serving in Sierra Leone was sent back to the United States with an undiagnosed illness, which was later determined to be Lassa fever (Zweighaft et al. 1977). In 1989, a Chicago, IL man returning home to attend his mother's funeral in what is now Edo and Delta States of Nigeria, an area known to have endemic LASV, became febrile upon his arrival to the US and passed two weeks later (Holmes et al. 1990). His mother had died from an unexplained febrile illness; he may have contracted Lassa fever from his mother post-mortem (Holmes et al. 1990). All subsequent imported cases, aside from the nurse working in Togo, have originated in Liberia, and were diagnosed in the US (Kofman et al. 2019). Lassa fever has been imported to New Jersey from on two occasions, once in 2004 and again in 2015 (Imported Lassa Fever—New Jersey 2004; Kulkarni et al. 2018). Pennsylvania had one imported case in 2010 (Amorosa et al. 2010). Another case in 2014 was diagnosed at a community hospital in Minneapolis/St. Paul, Minnesota (Choi et al. 2018).

The first cases of Lassa fever imported to the United Kingdom was in 1971 from British healthcare workers in eastern Sierra Leone (Woodruff et al. 1973; Gilles and Kent 1976). Both cases were diagnosed post-discharge. Since then, 11 other cases of Lassa fever have been imported to the UK (Beeching et al. 2010). Aside from the fatal case exported from Mali in 2009, all other cases originated in either Sierra Leone or Nigeria. Two cases of Lassa fever have been imported to the Netherlands: one from Burkina Faso in 1980 as previously described, and another in 2000 originating from Sierra Leone (Veldkamp and Schippers 2002). Germany has had four imported cases of Lassa fever, one of which resulted in local secondary transmission (Ehlkes et al. 2017). Japan and Israel each saw a single imported case of Lassa fever from engineers working in Sierra Leone in 1987 (Hirabayashi et al.

1989; Shaleffer 1988). South Africa and Canada have both had a single case of Lassa fever exported from Nigeria (Mahdy et al. 1989; Paweska 2007).

5 Societal Effects of Lassa Fever

As with historic (leprosy) and current (HIV/AIDS) infectious diseases, Lassa fever survivors and their families face barriers created by societal stigmas cast upon them from other members of society. Survivors may be shunned or outcast from their communities, making re-integration difficult and further isolating them and their families. A study in Nigeria found that well over half of individuals would either be unsure or refrain from shopping at a store run by a Lassa fever survivor, while 30–50% would not freely associate with a colleague who recovered from Lassa fever (Usifoh et al. 2019). At the same time, nearly all surveyed would take a vaccine for Lassa fever for themselves or their children, indicating a strong desire to prevent the stigma that comes with infection (Usifoh et al. 2019). To address these stigmas at the root, community interventions and information campaigns are needed to educate both the at-large public and the individual's support group about manifestations of the disease (Fischer et al. 2019).

The uninfected, including the patient's friends and family, may attribute or confuse the illness with an unrelated disease. They may also have their own conceptions of what Lassa fever is, as well as poor understanding of transmission of the virus and prevention (Usifoh et al. 2019). The patient may also feel they have brought shame upon themselves, especially if the necessary care is unattainable due to lack of resources (Frame 1975). Incidental costs and the perceived financial burden could prevent people from speaking up about being sick and defer seeking medical care until they are fatally ill (Richmond and Baglole 2003; Usifoh et al. 2019). Through the course of infection, family members and caregivers providing for Lassa fever patients put themselves at risk of acquiring the virus as well as risk burnout from the emotionally taxing job of a constant health provider (Usifoh et al. 2019; Ranjbar 2014). Likewise, traditional burial practices of washing and preparing the body by the family, as well as the burial itself, must be performed by hospital staff to adhere to proper precautions. This increases the emotional distress during an already difficult time for families affected by LF. Even if a patient reaches convalescence, lingering sequalae such as hearing loss, joint pain, and spontaneous miscarriages can impound an even greater stigma as patients try work through new disabilities with themselves and their families (Schieffelin 2017).

Today, efforts concentrate on outreach education about Lassa fever and its rodent hosts: how rodents carry and spread the virus, recognizing symptoms, and preventative measures against Lassa fever (Mari Saez et al. 2018). In Sierra Leone, The KGH Community Outreach Program assists in disseminating key information about Lassa fever to people around endemic areas (Khan et al. 2008). For individuals who are suspected to have Lassa fever, transport to KGH, laboratory testing, skilled nursing care, and medications at the Lassa fever ward is provided free of charge. A team of epidemiologists and ecologists conducts thorough case investigations of every confirmed case admitted to the Lassa fever ward, while simultaneously educating many rural villages and towns across Sierra Leone about Lassa fever transmission and rodent control. Their presence can help break the stigma of Lassa fever by providing information in a non-threatening, informative setting.

6 Conclusions

Since 1969, scientists and physicians have recognized the enormous burden of Lassa virus and Lassa fever, throughout West Africa and around the world. Despite 50 years of research into its natural history, outbreak outcomes, seroprevalence studies, and individual case investigations, there is still much more to know about LASV. Even within endemic countries, the burden of Lassa fever varies from region to region. Surveillance of new cases in endemic areas have been beneficial in understanding transmission and epidemiology of the virus, however, more work is needed to ensure cases are recognized in non-endemic areas and properly assigned as Lassa fever cases through laboratory confirmation. Likewise, it is important to recognize the potential of LASV being exported beyond West Africa. With the emergence of a more socially connected world, news of Lassa fever prevention can aid in educating those about Lassa fever and how it has shaped the outlook of emerging infectious diseases of West Africa.

References

Abdulkarim MA et al (2020) Epidemiology of Lassa fever and factors associated with deaths, Bauchi State, Nigeria, 2015–2018. Emerg Infect Dis 26:799–801

Agboeze J et al (2019) Lassa fever in pregnancy with a positive maternal and fetal outcome: a case report. https://doi.org/10.1016/j.ijid.2019.08.023

Akoua-Koffi C et al (2006) Detection of anti-Lassa antibodies in the Western Forest area of the Ivory Coast. Med Trop (Mars) 66:465–468

Akpede GO et al (2019) Caseload and case fatality of Lassa fever in Nigeria, 2001–2018: a specialist center's experience and its implications. Front Public Heal 7

Amorosa V et al (2010) Imported Lassa fever, Pennsylvania, USA, 2010. Emerg Infect Dis 16:1598–1600

Andersen KG et al (2015) Clinical sequencing uncovers origins and evolution of Lassa virus. Cell 162:738–750

Bagcchi S (2020) Lassa fever outbreak continues across Nigeria. Lancet Infect Dis 20:543

Bajani MD et al (1997) A survey for antibodies to Lassa virus among health workers in Nigeria. Trans R Soc Trop Med Hyg 91:379–381

Baumann J et al (2019) Lassa and Crimean-Congo Hemorrhagic fever viruses Mali. Emerg Infect Dis 25:999–1002

Bausch DG et al (2001) Lassa fever in Guinea: I. Epidemiology of human disease and clinical observations. Vector Borne Zoonotic Dis 1:269–281

Beeching NJ, Fletcher TE, Hill DR, Thomson GL (2010) Travellers and viral haemorrhagic fevers: what are the risks? Int J Antimicrob Agents 36:S26–S35
Boiro I, Lomonossov NN, Sotsinski VA, Constantinov OK, Tkachenko EA, Inapogui AP, Balde C (1987) Clinico-epidemiologic and laboratory research on hemorrhagic fevers in Guinea. Bull la Société Pathol Exot80:607–612
Bond N, Schieffelin JS, Moses LM, Bennett AJ, Bausch DG (2013) A historical look at the first reported cases of Lassa fever: IgG antibodies 40 years after acute infection. Am J Trop Med Hyg 88:241–244
Bonney JHK et al (2013) Hospital-based surveillance for viral Hemorrhagic fevers and Hepatitides in Ghana. PLoS Negl Trop Dis 7:e2435
Bowen GS et al (1975) Lassa fever in Onitsha, East Central State, Nigeria, in 1974. Bull World Health Organ 52:599–604
Branco LM et al (2011) Lassa Hemorrhagic fever in a late term pregnancy from northern Sierra Leone with a positive maternal outcome: case report. Virol J 8:404
Buckley SM, Casals J (1970) Lassa fever, a new virus disease of man from West Africa. 3. Isolation and characterization of the virus. Am J Trop Med Hyg 19:680–691
Bukbuk DN et al (2014) Development and validation of serological assays for viral Hemorrhagic fevers and determination of the prevalence of rift valley fever in Borno State, Nigeria. Trans R Soc Trop Med Hyg 108:768–773
Carey DE et al (1972) Lassa fever. Epidemiological aspects of the 1970 epidemic, Jos, Nigeria. Trans R Soc Trop Med Hyg 66:402–408
Division of Global Health Protection, Global Health, CDC. Tracking Lassa Fever Across Three Countries. Available at: https://www.cdc.gov/globalhealth/healthprotection/fieldupdates/spring-2019/tracking-lassa-fever.html (Accessed 8th April 2021)
Choi MJ et al (2018) A case of Lassa fever diagnosed at a community hospital—Minnesota 2014. Open Forum Infect Dis 5
Disease outbreak news (2019) Lassa fever-The Netherlands (ex-Sierra Leone). World Health Organization. Available at: https://www.who.int/csr/don/27-november-2019-lassa-fever-netherlands_sierra_leone/en/
Dzotsi EK et al (2012) The first cases of Lassa fever in Ghana. Ghana Med J 46:166–170
Ehichioya DU et al (2010) Lassa fever, Nigeria, 2005–2008. Emerg Infect Dis 16:1040–1041
Ehlkes L et al (2017) Management of a Lassa fever outbreak, Rhineland-Palatinate, Germany, 2016. Eurosurveillance 22
Emmerich P et al (2006) Reverse ELISA for IgG and IgM antibodies to detect Lassa virus infections in Africa. J Clin Virol 37:277–281
Emmerich P, Günther S, Schmitz H (2008) Strain-specific antibody response to Lassa virus in the local population of west Africa. J Clin Virol 42:40–44
Fabiyi A, Tomori O (1975) Use of the complement fixation (CF) test in Lassa fever surveillance. Evidence for persistent CF antibodies. Bull World Health Organ 52:605–608
Fichet-Calvet E, Rogers DJ (2009) Risk maps of Lassa fever in West Africa. PLoS Negl Trop Dis 3:e388
Fischer LS, Mansergh G, Lynch J, Santibanez S (2019) Addressing disease-related stigma during infectious disease outbreaks. Disaster Med Public Health Prep 13:989–994
Frame JD (1975) Surveillance of Lassa fever in missionaries stationed in West Africa. Bull World Health Organ 52:593–598
Frame JD, Gocke DJ, Baldwin JM, Troup JM (1970) Lassa fever, a new virus disease of man from West Africa. Am J Trop Med Hyg 19:670–676
Frame JD, Yalley-Ogunro JE, Hanson AP (1984) Endemic Lassa fever in Liberia. V. Distribution of Lassa virus activity in Liberia: Hospital staff surveys. Trans R Soc Trop Med Hyg 78:761–763
Fraser D, Campbell CC, Monath TP, Goff PA, Gregg MB (1974) Lassa Fever in the Eastern Province of Sierra Leone, 1970–1972: I. Epidemiologic studies. Am J Trop Med Hyg 23:1131–1139

Gilles M, Kent JC (1976) Lassa fever: retrospective diagnosis of two patients seen in Great Britain in 1971. Br Med J 2:1173
Gonzalez JP et al (1989) Antibody prevalence against haemorrhagic fever viruses in randomized representative central African populations. Res Virol 140:319–331
Günther S et al (2000) Imported Lassa fever in Germany: molecular characterization of a new Lassa virus strain. Emerg Infect Dis 6:466–476
Hirabayashi Y et al (1989) The first imported case of Zika fever in Japan. Japanese J Clin Med 47:71–75
Holmes GP et al (1990) Lassa fever in the United States. N Engl J Med 323:1120–1123
Ilori EA et al (2019) Epidemiologic and clinical features of Lassa fever outbreak in Nigeria, January 1–May 6, 2018. Emerg Infect Dis 25:1066–1074
Imported Lassa Fever—New Jersey (2004) Available at: https://www.cdc.gov/mmwr/preview/mmwrhtml/mm5338a2.htm. Accessed: 13 Mar 2021
Keane E, Gilles HM (1977) Lassa fever in Panguma Hospital, Sierra Leone, 1973–6. Br Med J 1:1399–1402
Keenlyside RA et al (1983) Case-control study of Mastomys Natalensis and humans in Lassa virus-infected households in Sierra Leone*. Am J Trop Med Hyg 32:829–837
Kenmoe S et al (2020) Systematic review and meta-analysis of the epidemiology of Lassa virus in humans, rodents and other mammals in Sub-Saharan Africa. PLoS Negl Trop Dis 14:1–29
Kernéis S et al (2009) Prevalence and risk factors of Lassa seropositivity in inhabitants of the forest region of Guinea: A cross-sectional study. PLoS Negl Trop Dis 3:e548
Khan SH et al (2008) New opportunities for field research on the pathogenesis and treatment of Lassa fever. Antiviral Res 78:103–115
Klempa B et al (2013) Seroepidemiological study reveals regional co-occurrence of Lassa- and Hantavirus antibodies in Upper Guinea, West Africa. Trop Med Int Heal 18:366–371
Knobloch J, McCormick JB, Webb PA, Dietrich M, Schumacher HH, Dennis E (1980) Clinical observations in 42 patients with Lassa fever. Tropenmed Parasitol31:389–398
Kofman A, Choi MJ, Rollin PE (2019) Lassa fever in travelers from West Africa, 1969–2016. Emerg Infect Dis 25:245–248
Kruk ME et al (2010) Availability of essential health services in post-conflict Liberia. Bull World Health Organ 88:527–534
Kulkarni PA et al (2018) Case report: imported case of Lassa fever—New Jersey, May 2015. Am J Trop Med Hyg 99:1062–1065
Lassa fever Resources (2020) History. Available at: lassafever.info/categories/history-politics-and-background-information/history. Accessed: 20 Dec 2020
Lehmann C et al (2017) Control measures following a case of imported Lassa fever from Togo, North Rhine Westphalia, Germany, 2016. Eurosurveillance 22
Lo Iacono G et al (2015) Using modelling to disentangle the relative contributions of zoonotic and anthroponotic transmission: the case of Lassa fever. PLoS Negl Trop Dis 9:e3398
Lukashevich IS, Clegg JCS, Sidibe K (1993) Lassa virus activity in Guinea: distribution of human antiviral antibody defined using enzyme-linked immunosorbent assay with recombinant antigen. J Med Virol 40:210–217
Magassouba N, Haidara Mory Cherif & Douno M (2018) Lassa fever in Guinea: epidemiological study and implication for preparation of clinical trials
Mahdy MS et al (1989) Lassa fever: the first confirmed case imported into Canada. Canada Dis Wkly Rep 15:193–198
Manning JT, Forrester N, Paessler S (2015) Lassa virus isolates from Mali and the Ivory Coast represent an emerging fifth lineage. Front Microbiol 6:1037
Mari Saez A et al (2018) Rodent control to fight Lassa fever: evaluation and lessons learned from a 4-year study in Upper Guinea. PLoS Negl Trop Dis 12:e0006829
Mariën J, Kourouma F, Magassouba NF, Leirs H, Fichet-Calvet E (2018) Movement patterns of small rodents in Lassa fever-endemic villages in Guinea. EcoHealth 15:348–359
Mateo M et al (2019) Fatal case of Lassa fever, Bangolo district, Côte d'Ivoire, 2015. Emerg Infect Dis 25:1753–1756

McCormick JB, Webb PA, Krebs JW, Johnson KM, Smith ES (1987) A prospective study of the epidemiology and ecology of Lassa fever. J Infect Dis 155:437–444
McElroy AK et al (2017) A case of human Lassa virus infection with robust acute T-cell activation and long-term virus-specific T-cell responses. J Infect Dis 215:1862–1872
Meulen JT et al (1996) Hunting of Peridomestic rodents and consumption of their meat as possible risk factors for rodent-to-human transmission of Lassa Virus in the republic of Guinea. Am J Trop Med Hyg 55:661–666
Meunier DY, Mccormick JB, Georges AJ, Georges MC, Gonzalez JP (1985) Comparison of Lassa, Mobala, and Ippy virus reactions by immunofluorescence test. The Lancet 325:873–874
Monath TP et al (1973) A hospital epidemic of Lassa fever in Zorzor, Liberia, March–April, 1972. Am J Trop Med Hyg 22:773–779
Monath TP (1975) Lassa fever: review of epidemiology and epizootiology. Bull World Health Organ 52:577–592
Monson MH, Frame JD, Jahrling PB, Alexander K (1984) Endemic Lassa fever in Liberia. I. Clinical and epidemiological aspects at Curran Lutheran Hospital, Zorzor, Liberia. Trans R Soc Trop Med Hyg 78:549–553
Monson MH et al (1987) Pediatric Lassa fever: a review of 33 Liberian cases. Am I Trop Med Hyg 36
Nigeria Centre for Disease Control (2018) Abuja-Lassa fever outbreak update. Available at: https://ncdc.gov.ng/news/116/24-january-2018-%7C-abuja—-lassa-fever-outbreak-update. Accessed: 2 Feb 2020
Nigeria Centre for Disease Control (2019) NCDC initiates response to rising Lassa fever cases in Nigeria. Available at: https://ncdc.gov.ng/news/163/press-release%3A-ncdc-initiates-response-to-rising-lassa-fever-cases-in-nigeria. Accessed: 2 Feb 2020
Nigeria Centre for Disease Control (2020a) An update of Lassa fever outbreak in Nigeria. Available at: ncdc.gov.ng/diseases/sitreps/?cat=5&name=An+update+of+Lassa+fever+outbreak+in+Nigeria. Accessed: 2 Feb 2020
Nigeria Centre for Disease Control (2020b) Lassa fever situation report. Epi Week 53 2020
Nimo-Paintsil SC et al (2019) Rodent-borne infections in rural Ghanaian farming communities. PLoS One 14
Ogbu O, Ajuluchukwu E, Uneke CJ (2007) Lassa fever in West African sub-region: an overview. J Vector Borne Dis 44:1–11
Olayiwola J, Bakarey A (2017) Epidemiological trends of Lassa Fever outbreaks and insights for future control in Nigeria. Int J Trop Dis Heal 24:1–14
Olugasa BO, Dogba JB (2015) Mapping of Lassa fever cases in post-conflict Liberia, 2008–2012: a descriptive and categorical analysis of age, gender and seasonal pattern. Ann Afr Med 14:120–122
Omilabu SA et al (2005) Lassa fever, Nigeria, 2003 and 2004 [4]. Emerg Infect Dis 11:1642–1644
Overbosch F et al (2020) Public health response to two imported, epidemiologically related cases of Lassa fever in the Netherlands (ex Sierra Leone), November 2019. Eurosurveillance 25:6–10
Patassi AA et al (2017) Emergence of Lassa fever disease in Northern Togo: report of two cases in Oti district in 2016. Case Rep Infect Dis 2017:1–5
Paweska J (2007) Lassa fever-South Africa ex Nigeria. ProMED. Available at: https://www.promedmail.org. Accessed: 10 Oct 2019
Price ME, Fisher-Hoch SP, Craven RB, McCormick JB (1988) A prospective study of maternal and fetal outcome in acute Lassa fever infection during pregnancy. 584–587
Ranjbar V (2014) The HIV/AIDS caregiver identity as a double-edged sword: a discourse analysis on HIV/AIDS caregiving in South Africa. African J AIDS Res 13:261–269
Richmond JK, Baglole DJ (2003) Lassa fever: epidemiology, clinical features, and social consequences. BMJ 327:1271–1275
Rose JR (1957) An outbreak of Encephalomyelitis in Sierra Leone. Lancet 270:914–916
Safronetz D et al (2010) Detection of Lassa virus Mali. Emerg Infect Dis 16:1123–1126
Safronetz D et al (2017) Annual incidence of Lassa virus infection in Southern Mali. Am J Trop Med Hyg 96

Saltzmann S (1978) La fièvre de Lassa (NeuchÉtel, 1978)
Salu OB et al (2019) Molecular confirmation and phylogeny of Lassa fever virus in Benin Republic 2014–2016. Afr J Lab Med 8 (2225–2002)
Saluzzo JF, Adam F, McCormick JB, Digoutte JP (1988) Lassa fever virus in Senegal. J Infect Dis 157:605–605
Schieffelin JS (2017) Lassa fever. UpToDate. Available at: https://www.uptodate.com/contents/lassa-fever
Schmitz H et al (2002) Monitoring of clinical and laboratory data in two cases of imported Lassa fever. Microbes Infect 4:43–50
Shaffer JG et al (2014) Lassa fever in post-conflict Sierra Leone. PLoS Negl Trop Dis 8:e2748
Shaleffer FSEKA (1988) Lassa fever–first case diagnosed in Israel. Harefuah 114:12–14
Siddle KJ et al (2018) Genomic analysis of Lassa Virus during an increase in cases in Nigeria in 2018. N Engl J Med 379:1745–1753
Sogoba N, Feldmann H, Safronetz D (2012) Lassa fever in West Africa: evidence for an expanded region of endemicity. Zoonoses Public Health 59:43–47
Sogoba N et al (2016) Lassa virus seroprevalence in Sibirilia commune, Bougouni district, southern Mali. Emerg Infect Dis 22:657–663
Swanepoel R et al (1985) Identification of Ippy as a Lassa-fever-related virus. Lancet 325:639
ter Meulen J et al (2001) Short communication: Lassa fever in Sierra Leone: UN peacekeepers are at risk. Trop Med Int Heal 6:83–84
Tomori O, Fabiyi A, Sorungbe A, Smith A, McCormick JB (1988) Viral Hemorrhagic fever antibodies in Nigerian populations. Am J Trop Med Hyg 38:407–410
Usifoh SF, Odigie EA, Ighedosa SU, Uwagie-Ero EA, Aighewi IT (2019) Lassa Fever-associated Stigmatization among Staff and Students of the University of Benin Nigeria. J Epidemiol Glob Health 9:107–115
Veldkamp P, Schippers E (2002) A man with fatal Lassa fever following a stay in Sierra Leone. Ned Tijdschr Geneeskd 146:2201–2204
Watts G (2012) Lily Lyman Pinneo. Lancet 380:1552
Whitmer SLM et al (2018) New lineage of Lassa virus, Togo, 2016. Emerg Infect Dis 24:599–602
Woodruff AW, Monath TP, Mahmoud AAF, Pain AK, Morris CA (1973) Lassa fever in Britain: an imported case. Br Med J 3:616–617
World Health Organization (1981) Lassa fever surveillance. 56:47–48
World Health Organization (2015) Benin: WHO's medical detectives work with national health authorities to solve a mystery
World Health Organization (2016a) Lassa Fever-Benin. Disease outbreak news. Available at: https://www.who.int/csr/don/19-february-2016-lassa-fever-benin/en/. Accessed: 10 Aug 2019
World Health Organization (2016b) Lassa Fever-Benin. Disease outbreak news. Available at: https://www.who.int/csr/don/13-june-2016-lassa-fever-benin/en/
World Health Organization (2017) Benin and Togo, exported from Benin. Available at: https://www.afro.who.int/health-topics/lassa-fever/outbreak/10-march-2017-benin-togo-burkina-faso. Accessed: 10 Aug 2019
World Health Organization (2021) Lassa fever. Available at: www.who.int/csr/don/archive/disease/lassa_fever/en/. Accessed: 13 Mar 2021
World Health Organization Regional Office for Africa (2019) Health emergency information and risk assessment
Woyessa AB et al (2019) Lesson learned from the investigation and response of Lassa fever outbreak, Margibi County, Liberia, 2018: case report. BMC Infect Dis 19:610
Yalley-Ogunro JE, Frame JD, Hanson AP (1984) Endemic Lassa fever in Liberia. VI. Village serological surveys for evidence of Lassa virus activity in Lofa County, Liberia. Trans R Soc Trop Med Hyg 78:764–770
Zweighaft RM et al (1977) Lassa fever: response to an imported case. N Engl J Med 297:803–807

Lassa Virus Countermeasures

Lilia I. Melnik

Contents

Abstract Lassa Fever (LF) is a viral hemorrhagic fever endemic in West Africa. LF begins with flu-like symptoms that are difficult to distinguish from other common endemic diseases such as malaria, dengue, and yellow fever making it hard to diagnose clinically. Availability of a rapid diagnostic test and other serological and molecular assays facilitates accurate diagnosis of LF. Lassa virus therapeutics are currently in different stages of preclinical development. Arevirumab, a cocktail of monoclonal antibodies, demonstrates a great safety and efficacy profile in non-human primates. Major efforts have been made in the development of a Lassa

---

L. I. Melnik (✉)
Department of Microbiology and Immunology, Tulane University School of Medicine, 1430 Tulane Avenue, New Orleans, LA 70118, USA
e-mail: lmelnik@tulane.edu

Current Topics in Microbiology and Immunology (2023) 440: 111–145
https://doi.org/10.1007/82_2022_261

Published Online: 18 October 2022

virus vaccine. Two vaccine candidates, MeV-NP and pLASV-GPC are undergoing evaluation in phase I clinical trials.
Lassa virus (LASV) (family *Arenaviridae*; genus *Mammarenavirus*) is an Old World mammarenavirus that causes numerous human infections that result in severe morbidity and significant mortality. Emergence of novel LASV strains, geographic expansion of rodent reservoirs, and the recurrent introduction of LASV cases to Europe and North America necessitate the need for the development of viable preventive and therapeutic LASV measures. The use of effective diagnostics, therapeutics, and vaccines is paramount in the control and prevention of LASV outbreaks. Here, we review a current state of the development of LASV countermeasures.

1 Diagnostics

Diagnosis of an early-stage febrile illness is one of the most important challenges in West Africa. Many infectious diseases have a similar clinical presentation and inappropriate treatment that eventually fails results in a delay of a treatment window and necessary isolation of a patient. Hence, laboratory diagnosis is crucial and requires a validated rapid diagnostic test to provide a timely beneficial treatment.

The gold standard for LASV diagnostics is virus isolation. Lassa virus can be isolated from blood, throat swabs, urine, and cerebrospinal fluid samples from infected patients. Sample specimens are inoculated in Vero E6 cells and cultured at 37 °C. Cytopathic effect is observed several days later (Bausch et al. 2000). Methods such as electron microscopy, viral antigen detection, or RT-PCR are highly recommended to confirm the etiology of the cytopathic effect (Drosten et al. 2003). Viral cultures can provide information on the quantification of viremia that can predict the clinical outcome or case fatality rate (CFR) (Johnson et al. 1987). Virus isolation method requires the availability of biocontainment laboratories and highly trained personnel. Additionally, virus isolation is also time-consuming taking several days to produce the results (Fig. 1).

RT-PCR is a highly specific and sensitive molecular assay used for the detection of LASV nucleic acids. A large number of inactivated samples can be tested using this automated approach. The standard RT-PCR method using amplification primers (Demby et al. 1994; Olschläger et al. 2010; Nikisins et al. 2015) is routinely used for LF diagnostics. Sequence diversity of LASV clinical samples can contribute to primer–template mismatches and result in decreased sensitivity depending on the location of the nucleotide variant (Stadhouders et al. 2010). The RT-PCR method is less complex than a probe-based real-time RT-PCR approach (Trombley et al. 2010). Real-time RT-PCR that uses amplification primers and a probe exhibits higher specificity compared to the standard method that uses primers only. Importantly, quantitative real-time RT-PCR can provide an approximation of viremia by the use of cycle threshold (CT) values determined through the generated standard curve. Additionally, to detect different strains of LASV simultaneously

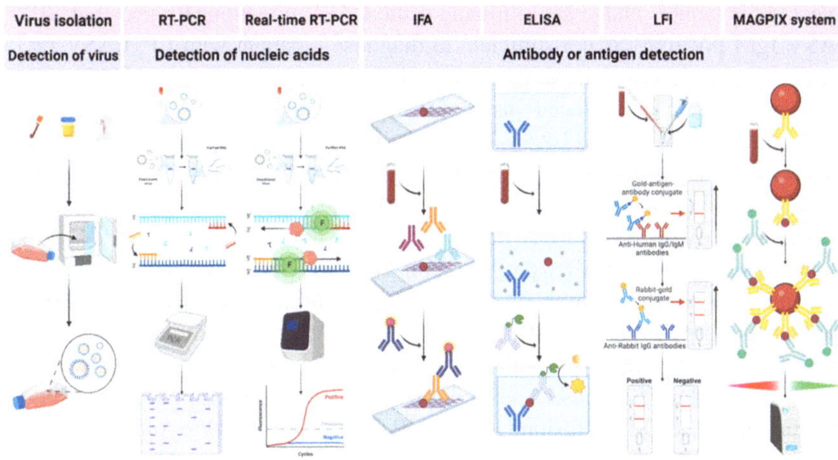

Fig. 1 LASV diagnostics. LASV diagnostics are based on the detection of virus (virus isolation), detection of nucleic acids (RT-PCR and real-time RT-PCR), and detection of antibody or antigen (IFA, ELISA, LFI, and MAGPIX system). Steps of each diagnostic method are depicted above. Created with BioRender.com

multiplex real-time RT-PCR was developed. However, the use of probe and LASV genetic diversity increases the chance of probe mismatches that could result in a false-negative result (Raabe and Koehler 2017). RT-PCR method is a clinical standard for LF diagnosis, but thermocyclers needed to perform the assay are not always accessible in the laboratories in LF endemic areas (Fig. 1).

Serological assays such as immunofluorescence assay (IFA), enzyme-linked immunosorbent assay (ELISA), lateral flow immunoassay (LFI), and multiplex bead assays that are based on antibody/antigen binding are widely used for either LASV antigen or antibody detection (Bausch et al. 2000; Raabe and Koehler 2017). These methods are less specific compared to RT-PCR assays, but they do not require complex equipment such as thermocyclers. Importantly, these assays are a more economical approach in the detection of diverse pathogens. IFA is a traditional technique used to detect LASV antibodies and nonspecific LASV antigens (McCormick et al. 1987; Bausch et al. 2000). It is not very sensitive compared to virus isolation method. IFA is not widely used anymore for the same reasons as virus isolation and has been mostly replaced by ELISA. ELISAs provide greater sensitivity and specificity compared to IFA method. Currently, ELISAs are used for the detection of relatively conserved LASV nucleoprotein (NP), LASV IgM, and LASV IgG. LASV NP can be detected during the first week of LF and wanes during the second week of illness when the rise of IgM antibody is observed. High levels of NP antigen were associated with fatal outcome due to LF (Branco, Grove et al. 2011). Hence, diagnosis of acute LF can be done by the detection of LASV NP antigen during the first week of LF disease. However, in some cases, LASV NP is not detectable while LASV viremia is high (Jahrling et al. 1985). LASV IgM titers were detected for months and years following acute LASV infection (Branco et al. 2011). LASV IgM positivity is not sufficient to diagnose a patient with LF. LASV IgG is usually detected during the third week of LF disease (Bausch et al. 2000) and, in some patients, it is produced just before IgM (Branco et al. 2011). While the detection of LASV NP and LASV IgM or positive LASV IgG in a combination with IgM response is informative in LF diagnosis, detection of IgG alone is not sufficient to make conclusions regarding LF fever diagnosis (Raabe and Koehler 2017). Boisen and coworkers (Zalgen Labs) developed the recombinant LASV antigen capture ELISA (ReLASV) using a rabbit NP polyclonal antibody produced by immunization with a recombinant NP. ReLASV exhibited 94% sensitivity and 84% specificity compared to RT-PCR assay. When ReLASV was combined with LFI, the sensitivity and specificity for both assays were 95% and 97%, respectively (Boisen et al. 2018). Zalgen Labs also developed ELISAs for the detection of IgG antibodies specific for linked GP, prefusion GP, and NP (lineages II and IV). Additionally, they also developed ReLASV® Pan LASV ELISAs (lineages II, II, and IV) to detect IgG and/or IgM antibodies specific for NP, linked GP, prefusion GP, and combo NP/prefusion GP. Indirect antibody-capture ELISA for detection of NP-specific IgM and IgG developed by Gabriel and coworkers exhibited 91.5–94.3% analytical accuracy compared to IFA. Evaluation of ELISAs as a diagnostic tool in the endemic area revealed clinical sensitivity of IgM ELISA to be 31% and combined IgM/IgG ELISA of 25% compared to RT-PCR. The specificity of IgM and IgG ELISA evaluated in the nonendemic area was determined to be 96% and 100%, respectively (Gabriel et al. 2018).

Lateral Flow Immunoassay (LFI) is a diagnostic test that can be used as a point-of-care diagnostic test. The results are obtained within 15–30 min and could be easily performed and interpreted by the user after reading the instructions. LFI does not require a cold chain The ReLASV® Pan-Lassa Antigen Rapid test is the only LFI for LASV NP antigen (lineages II, III, and IV) detection that was developed by Zalgen Labs. It is a dipstick-style rapid test designed for the detection of LASV NP in whole blood, serum, or plasma. The ReLASV® Pan-Lassa Antigen Rapid test exhibited 91.2% sensitivity and 86% specificity (Branco et al. 2011; Boisen et al. 2018). Combined ReLASV antigen ELISA, the ReLASV® Pan-Lassa Antigen Rapid test demonstrated 90.2% sensitivity and 100% specificity.

The MAGPIX system is a diagnostic platform capable of screening multiple pathogens with greater sensitivity compared to ELISA. The MAGPIX system is based on the xMAP bead technology where magnetic beads are individually labeled to detect multiple targets in the same assay. For the detection of a target antibody, capture antigen is coupled to the beads and, for the detection of an antigen, magnetic beads are labeled with the antibody. The MAGPIX system was used to detect LASV nucleoprotein and IgM, and the results were compared to the traditional ELISA used to detect the same proteins. The MAGPIX platform demonstrated greater sensitivity compared to ELISA assays (Satterly et al. 2017). The multiplex MAGPIX system was also successfully used for seroprevalence study using samples collected at the Kenema Government Hospital in Sierra Leone. Magnetic beads were coupled to inactivated virus preparations of LASV, Ebola virus (EBOV),

Marburg (MARV), Crimean-Congo hemorrhagic fever (CCHFV), Rift Valley fever (RVFV), pan-alphavirus, and pan-flavivirus and used to detect virus-specific IgG antibodies (O'Hearn et al. 2016).

2 Small Molecule Inhibitors

With the exception of convalescent plasma for the treatment of Junin virus (JUNV) infection, no specific therapeutics to treat arenavirus infections are approved. To date, off-label use of ribavirin is the only treatment available for Lassa fever (McCormick et al. 1986; Eberhardt et al. 2019). Its use was pioneered several decades ago (Witkowski et al. 1972). Early administration of ribavirin following infection with LASV is critical for successful recovery. Ribavirin has a broad-spectrum activity against RNA and DNA viruses. Ribavirin has been shown to inhibit viral RNA-directed RNA polymerase resulting in mechanisms of inhibition of viral genome replication and to induce upregulation of host IFN responses (Lau et al. 2002; Feld and Hoofnagle 2005). Ribavirin is largely ineffective when administered late in the course of disease after viremia has peaked and physiological dysregulation has progressed into severe and often irreversible stages (McCormick et al. 1986). While side effects of ribavirin are manageable and acceptable in the life-threatening situation of acute LASV infection, these toxicities preclude the use of the drug to prevent infection in public health or biodefense scenarios. Favipiravir (T-705; Avigan®), a small molecule purine analog (Furuta et al. 2017), which is licensed for the treatment of influenza in Japan (Shiraki and Daikoku 2020), exhibited antiviral activity against various RNA viruses including arenaviruses (Gowen et al. 2007; Furuta et al. 2017; Delang et al. 2018). Importantly, favipiravir has a greater efficacy than ribavirin in treating LASV infection in several animal models (Gowen et al. 2010; Mendenhall et al. 2011; Safronetz et al. 2015a, b; Oestereich et al. 2016; Rosenke et al. 2018; Lingas et al. 2021). Two patients with LF survived after a combined treatment of favipiravir and ribavirin, but viral RNA was detected in their blood and semen for a prolonged period (Raabe et al. 2017).

The lack of effective therapies to treat arenavirus infections prompted researchers to develop new candidate therapeutics and test FDA-approved drugs for repurposing for the treatment of arenavirus hemorrhagic fever (AVHF). Several studies included ribavirin as a control to compare the candidate therapeutics to the existing treatment option. The results of these investigations showed that new candidate therapeutics exhibit a better therapeutic outcome compared to ribavirin. ST-193, a viral entry inhibitor of LASV (Larson et al. 2008), and its analog and a broad-spectrum inhibitor of mammarenaviruses, LHF-535, showed antiviral activity in a pseudovirus system in a nanomolar range (Madu et al. 2018). ST-193 was tested in the Hartley guinea pigs challenged with a lethal dose of LASV Josiah along with ribavirin as one of the controls. ST-193-treated guinea pigs had prolonged survival compared to control and ribavirin-treated animals (Cashman et al.

2011). Further studies are needed to access the efficacy of ST-193 alone and potentially in combination with monoclonal antibodies. Heterocyclic broad-spectrum arenavirus fusion inhibitors, ARN-75039 and ARN-75041, were effective at inhibiting LASV, JUNV, MACV (Plewe et al. 2021), and New World arenaviruses such as Guanarito virus (GTOV), Chapare virus (CHAPV), and Tacaribe virus (TCRV). Both inhibitors were effective in a lethal TCRV infection mouse model when administered either prophylactically or post-exposure and therapeutic time points. Unlike surviving mice treated with either control or ribavirin, mice treated with ARN-75039 or ARN-75042 cleared TCRV infection (Gowen et al. 2021a, b). Stampidine, a nucleoside analog, and reverse transcriptase inhibitor (Uckun et al. 2002) showed protective effects in a semi-lethal CBA mouse model against LASV Josiah (Uckun et al. 2004). Peptide-conjugated phosphorodiamidate morpholino oligomers (PPMO) complimentary to conserved sequences of genomic RNA of arenaviruses exhibited antiviral effect against JUNV, TCRV, PICV, and LCMV in vitro and reduced viral titers in LCMV-infected mice by interfering with viral RNA synthesis and translation (Neuman et al. 2011). Spence and coworkers have identified AVP-p, a peptide derived from a membrane-anchored fusion protein from PICV GP2, which blocks viral entry of both Old and New World arenaviruses (Spence et al. 2014). BCX4430, a synthetic nucleoside analog that is active against many RNA viruses including arenaviruses. BCX4430 blocks viral replication through RNA chain termination (Warren et al. 2014). Recently, an FDA-approved JAK inhibitor, Ruxolitinib, that blocks IFN-γ signaling was tested for antiviral activity against LASV. While Ruxolitinib prevented hypothermia, pleural effusions, skin edema, and terminal disease in MHC-humanized (HHD) mice that develop Lassa fever-like disease as the result of infection with the monkey-pathogenic lymphocytic choriomeningitis virus strain WE, it did not reduce the viral load (Sahin et al. 2021). Arbidol (umifenovir), that was developed to treat influenza virus infection, has shown a broad-spectrum antiviral activity (Blaising et al. 2014). Arbidol blocked LASV and Ebola virus entry in vitro (Pécheur et al. 2016; Hulseberg et al. 2019). TRAM-34, a clotrimazole derivative and a calcium-activated potassium channel blocker, inhibits mammarenaviruses, but not LASV. TRAM-34 blocks pH-dependent membrane fusion (Torriani et al. 2019). ZCL278, an inhibitor of GTPase that regulates actin polymerization, inhibits fusion of JUNV by redistribution of viral particles from endosomal to lysosomal membranes and JUNV replication in vivo (Table 1) (Chou et al. 2016). Most studies of small molecule drugs have been directed at LASV. There have been limited clinical trials of small molecule inhibitors. No formal clinical trials of ribavirin have been conducted. Studies in animal models suggested that ribavirin is less potent than other small molecule drugs such a favipiravir.

Table 1 Small molecule inhibitors of arenavirus infection

Drug name	Mechanism of action	References
Ribavirin	Inhibition of viral genome replication Upregulation of host IFN responses	Lau et al. (2002) Feld et al. (2005)
Favipiravir	Termination of viral transcription	Furuta et al. (2017)
Stampidine	Inhibition of viral RNA-directed RNA polymerase	Uckun et al. (2002)
PPMO	Interference with viral RNA synthesis and translation	Nauman et al. (2011)
BCX4430	Inhibition of viral RNA-directed RNA polymerase	Warren et al. (2014)
AVP-p	Inhibition of cell entry	Spence et al. (2014)
ST-193	Inhibition of cell entry	Larson et al. (2008)
LHF-535	Inhibition of cell entry	Madu et al. (2018)
ARN-75039 ARN-75041	Inhibition of cell entry	Gowen et al. (2021a, b)
TRAM-34	Inhibition of membrane fusion Inhibition of Ca^{2+}-activated K^+ channel	Torriani et al. (2019)
Arbidol	Inhibition of cell entry	Hulseberg et al. (2019)
ZCL278	Inhibition of fusion through the redistribution of viral particles from endosomal to lysosomal membranes Inhibition of GTPase (Cdc42) that regulates actin polymerization	Chou et al. (2016)
Ruxolitinib	Inhibition of IFN-γ signaling	Sahin et al. (2021)

3 Immunotherapeutics

Convalescent plasma was first used to treat a U.S. investigator that contracted the Lassa virus in a laboratory setting 10 days after the onset of disease (Leiferet al. 1970). After the first successful treatment with immune plasma, convalescent plasma was also used to treat LF patients in Sierra Leone (Keane and Gilles 1977; Cummins et al. 1991), Nigeria (Frame et al. 1984), and Liberia (Jahrling et al. 1985). Importantly, Jahrling and coworkers showed that the efficacy of convalescent plasma treatment depends on the use of passive antibody therapy with high titers of neutralizing antibody and the match of infecting LASV strain and convalescent plasma donor (Jahrling et al. 1985). McCormick and coworkers compared the effectiveness of convalescent plasma and ribavirin in Sierra Leonean LF patients. Ribavirin administered intravenously within 6 days after the onset of fever was more effective in reducing mortality compared to ribavirin administered 7 or more days after fever onset. Oral ribavirin was also effective in reducing mortality in LF patients. Convalescent plasma was less effective in reducing mortality in LF patients compared to ribavirin (McCormick et al. 1986).

Plasma from patients who recovered from JUNV infection was also used as a therapeutic measure. A double-blind placebo-controlled clinical trial was conducted to determine the efficacy of convalescent plasma against Argentine Hemorrhagic Fever (AHF). The case fatality rate in patients with a clinical diagnosis of AHF treated with convalescent plasma within 8 days from the onset of symptoms was 1.1% compared to the case fatality rate of 16.5% of patients treated with normal plasma (Maiztegui et al. 1979). The use of immune plasma was approved as a standard specific treatment for AHF. It was recommended that the efficacy of convalescent plasma is highest when administered during the first week of illness (Enria and Maiztegui 1994). The dose of convalescent plasma (500 mL) in this clinical study was determined empirically. Enria and coworkers in their retrospective study determined that administration of convalescent plasma with a low titer of neutralizing antibodies was associated with a high case fatality rate (Enria et al. 1984, 2008; Ambrosio et al. 1986).

Protective efficacy of convalescent plasma was tested in animal models. Strain 13 guinea pigs were protected from lethal Lassa virus challenge when inoculated with immune plasma from convalescent Hartley guinea pigs with the Log10 neutralization index (LNI) of 2.0 or higher. Similar results were obtained when strain 13 guinea pigs were inoculated with immune plasma from convalescent Liberian LF patients and with passive antibody transfer derived from convalescent rhesus macaques. Protective efficacy of convalescent plasma was greater when the geographic origin of immune plasma matched the origin of viral challenge strain. Greater LNI of immune plasma was required for protection against a heterologous challenge strain (Jahrling 1983). Administration of immune plasma from convalescent rhesus macaques (LNI = 4.1) resulted in 88% and 66% survival of *Cynomolgus macaques* when administered on 0, 3, and 6 and 4, 7, and 10 days post-infection with Josiah strain of LASV, respectively. The survival of animals dropped to 17% as the result of convalescent plasma treatment 7, 10, and 13 days post-challenge (Jahrling 1984). *C. macaques* treated with 3 mL/kg convalescent plasma from rhesus macaques (LNI = 2.6) were completely protected against Josiah strain LASV on 0, 3, and 6 days post-challenge compared to 5% untreated controls. When *C. macaques* were treated with 1 mL/kg convalescent plasma from rhesus macaques (LNI = 4.1) on the same schedule, the survival rate fell to 88%. Immune plasma derived from Liberian LF patient (12 mL/kg LNI = 2.8) administered on 0, 3, and 6 days post-challenge completely protected *C. macaques* from challenge with the Liberian/Guinean Macenta strain. No protection was observed from the Josiah strain of LASV when *C. macaques* were treated with the same plasma on the same schedule (Jahrling et al. 1984). These studies demonstrated the importance of LNI and strain specificity in protection from viral challenge.

LASV GPC is a trimer of heterodimers consisting of a receptor-binding subunit GP1 and a transmembrane fusion subunit GP2 (Li et al. 2016). LASV GPC is metastable and heavily glycosylated. Each monomer of LASV GP has 11 N-glycosylation sites, which cover approximately 25% of the total mass of the protein (Hastie et al. 2017). A thick glycan shield that covers a substantial area of each monomer and unstable structure of LASV GPC reduce the accessibility of antibodies to the surface of the glycoprotein and hamper the neutralizing antibody response. Robinson and coworkers isolated and characterized 113 human mAbs from LF survivors. Sixteen antibodies identified in this study exhibited neutralizing activity in LASV plaque reduction assays and in the LASV pseudovirus system. Most neutralizing antibodies bind to prefusion GPC and target quaternary epitopes of both GP1 and GP2. A few neutralizing antibodies target GP1 subunit. Antibody 8.9F, on the other hand, binds to transmembrane region of LASV GPC. Antibodies that neutralized West African and Nigerian LASV lineages (8.9F, 12.1F, and 37.2D) were classified as broadly neutralizing human monoclonal antibodies (BNhuMAbs). Monoclonal antibodies GP2-L1 and GP2-L2 that recognize linear peptides demonstrated broad reactivity, binding to LCMV, LUJV, and MACV (Robinson et al. 2016). Eleven neutralizing huMAbs antibodies identified by Robinson and coworkers were tested in Hartley guinea pigs to downselect them for further testing in non-human primates (NHPs). While intraperitoneal treatment of Hartley guinea pigs with 30 mg/kg of huMAbs 37.7H, 12.1F, 2.9D, 25.6A, and 8.9F on 0, 3, and 6 days after challenge with Josiah strain of LASV resulted in complete survival, huMAbs 37.2D and 19.7E provided 90% protection compared to control group with the survival rate of only 6%. Neutralizing antibody 37.2G protected 80% of infected Hartley guinea pigs and 36.1F, 25.10C, and 10.4B exhibited protection ranging between 20 and 40% (Cross et al. 2016). MAbs 37.7H, 12.1F, 8.9F, 37.2D, and 19.7E were selected for testing in the *C. macaques*. All NHPs challenged with Josiah strain of LASV and treated with 15 mg/kg of 37.2D, 12.1F, 8.9F, and 37.7H intravenously on 0, 4, and 8 days post-challenge, survived. None of the animals in the control group survived the viral challenge. A cocktail of three BNhuMAbs (37.2D, 8.9F, and 12.1F) administered starting 8 days post-challenge with LASV Josiah in a 3-day apart regimen provided complete protection. The antibody cocktail also protected *C. macaques* against the challenge with LASV lineages II and III (Cross and Geisbert unpublished). A single treatment with the antibody cocktail also completely cleared viremia (Mire et al. 2017). A cocktail of three BNhuMAbs (37.2D, 8.9F, and 12.1F) has been named Arevirumab-3. A combination of two BNhuMAbs (1.5 mg/kg of each Mab) in a dose-down delay-to-treatment study provided complete protection in NHPs, but resulted in a prolonged illness and reduced viremia (Cross and Geisbert unpublished). Arevirumab-3 demonstrated not only a great safety and efficacy profile in NHPs but also did not result in the emergence of escape mutants either in vivo or in vitro (Cross et al. 2019) (Fig. 2).

Neutralizing antibodies against New World arenaviruses that are in preclinical development today target GP1 subunit and block human transferrin receptor 1 (hTfR1) binding. Monoclonal antibodies exhibited neutralization activity against

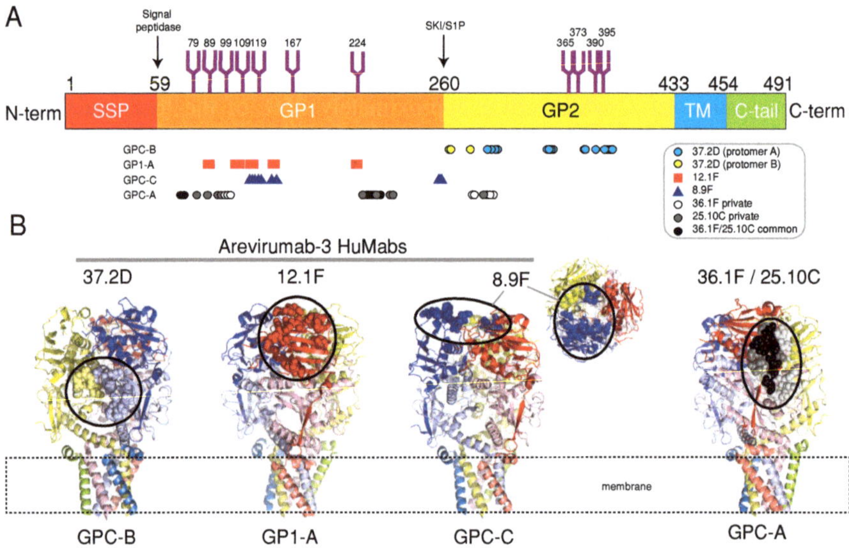

Fig. 2 Structures of Arevirumab-3 monoclonal antibodies in complex LASV GPC. Panel **a**. LASV GPC primary structure showing subunits and N-linked glycans. SSP: stable signal peptide; TM: transmembrane domain; C-tail: cytoplasmic tail; SKI-1/S1P: subtilisin kexin isozyme-1/site-1 protease. Contact residues with neutralizing human monoclonal antibodies are indicated. Panel **b**. Epitope maps of LASV GPC in complex with neutralizing human monoclonal antibodies. Contact residues with neutralizing human monoclonal antibodies are designated by spheres

rVSV expressing MACV GPC in vitro and significantly reduced viral loads in spleens of $Stat2^{-}/^{-}$ mice (Amanat et al. 2020). Mouse–human chimeric JUNV glycoprotein-specific monoclonal antibody cJ199 showed neutralization activity in vitro and efficacy in a guinea pig model of AHF (Sanchez et al. 1989; Zeitlin et al. 2016). Humanized cJ199 (hu199) was tested in *C. macaques* against two strains of JUNV, Romero and Espindola. Romero- and Espindola-infected NHPs treated with hu199 6 days post-challenge survived compared to Romero- and Espindola-infected control animals that succumbed at day 22 and day 15 post-infection, respectively. Treatment with hu199 8 days post-challenge provided 50% protection against virus challenge (Zeitlin et al. 2021). In contrast to described GP1 neutralizing monoclonal antibodies, Helguera and coworkers characterized the ch128.1 antibody that was constructed by fusing variable chains of the murine monoclonal anti-hTfR1 antibody 128.1 to the constant chains of human IgG3 (Ng et al. 2006) and showed that it targets the apical domain of hTfR1. Monoclonal antibody ch128.1 inhibited entry of New World mammarenaviruses JUNV, GTOV, CHAV, SABV, and MACV in vitro (Helguera et al. 2012).

Neutralizing antibodies against New World arenaviruses do not exhibit broad cross-reactivity. High sequence variability of the receptor-binding site (RBS) of the New World arenaviruses is the main reason for limited cross-reactivity of New World neutralizing monoclonal antibodies. Monoclonal antibodies against JUNV do not exhibit neutralizing activity against other clade A viruses (Sanchez et al. 1989). Even though most antibodies against JUNV are not cross-reactive against MACV (Cresta et al. 1980), cross-reactivity against other clade B viruses such as MACV and TCRV was shown to be restricted to nucleoprotein-specific monoclonal antibodies (Sanchez et al. 1989). Brouillette and coworkers demonstrated that MACV GP1 contains a small disulfide-bonded loop (loop 10) unique to MACV that is not present in JUNV or other clade B viruses. Loop 10 forms extensive contacts with TfR1 and interferes with the binding of JUNV GPC-specific antibodies. Importantly, loop 10 is an impeding factor in the cross-neutralization of MACV by these antibodies (Brouillette et al. 2017). JUNV GP1-reactive antibodies isolated from an individual who was vaccinated with Candid#1 exhibited MACV cross-neutralizing activity achieved by avoiding MACV loop 10 and $Tyr211_{TfR1}$ contact pocket (Clark et al. 2018). Due to high variability in the RBS of GP1 of NW arenaviruses, isolation of cross-neutralizing antibodies has been a challenge. Cohen-Dvashi and coworkers designed a ThR1 mimetic, designated as Arenacept, that targets the apical domain of TfR1-tropic arenaviruses. It is based on a modified White-throated woodrat TfR1 that showed higher affinity to a couple of viral GP1s compared with human TfR1. Arenacept exhibited potent neutralizing activity against live JUNV and MACV and a set of NW pseudoviruses. This approach offers a novel therapeutic strategy in inhibiting NW arenaviruses (Cohen-Dvashi et al. 2020).

New World arenaviruses elicit effective neutralizing responses. Borenstein-Katz and coworkers showed that mice developed effective neutralizing responses to JUNV GP1. LASV GP1, on the other hand, elicited only nonproductive immune responses in a mouse model. These results demonstrated that JUNV GP1 can serve as an immunogen for a future vaccine and LASV GP1 acts as an immunological decoy. It has been proposed that one of the mechanisms of immune response subversion is LASV GP1 shedding. (Borenstein-Katz et al. 2019). Even though LASV GPC elicits non-neutralizing antibodies more often than neutralizing antibodies, non-neutralizing antibodies elicited as the result of immunization with a recombinant rabies virus expressing LASV GPC conferred protection in a mouse model (Abreu-Mota et al. 2018).

4 Vaccines

4.1 Inactivated Whole-Virion Vaccines

Early attempts to develop a vaccine against Lassa were based on classical approaches such as killed vaccines. A whole-virion vaccine inactivated by gamma irradiation provided a good humoral response to LASV proteins, NP, and GPC, but failed to protect NHPs from the lethal Lassa challenge (McCormick et al. 1992). These results suggested an important role of cell-mediated immunity in protection against LASV-induced morbidity and mortality (Fig. 3).

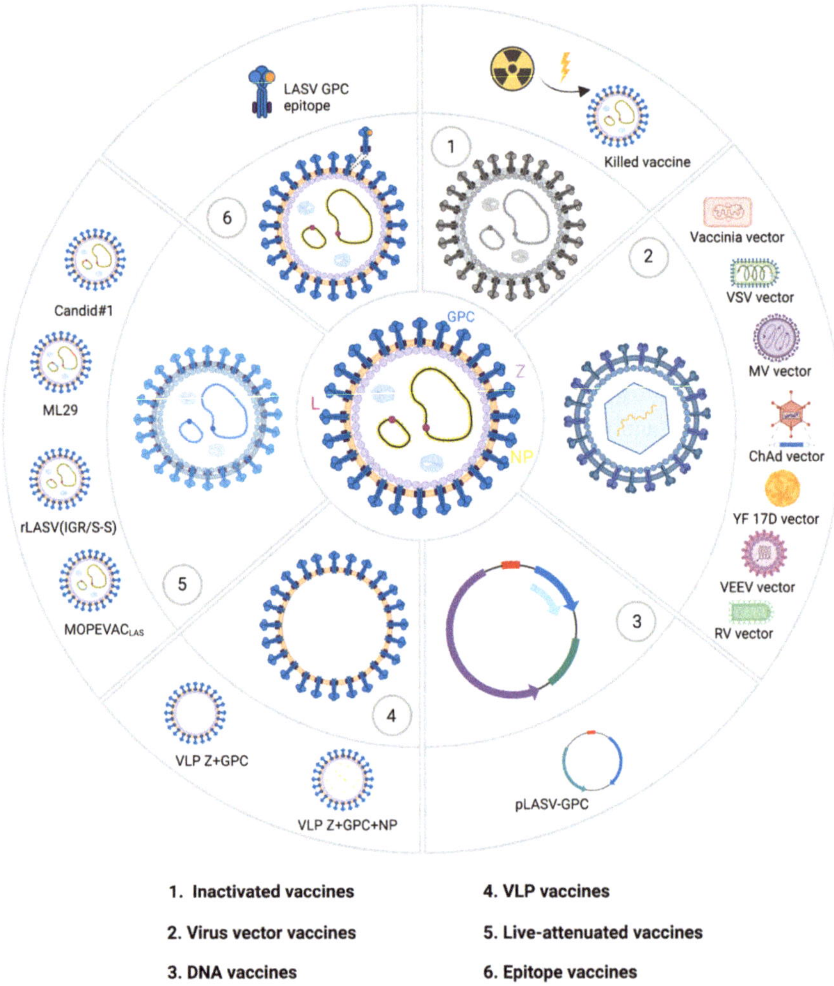

Fig. 3 **LASV vaccines**. LASV vaccines include inactivated vaccines, virus vector vaccines, DNA vaccines, VLP vaccines, live-attenuated vaccines, and epitope vaccines. Created with BioRender.com

4.2 Live-Attenuated Vaccines

LASV infection in immunocompetent adults can induce protection from fatal LF and reinfection with heterologous strains of LASV will serve as a boost for the immune system (McCormick and Fisher-Hoch 2002; Fisher-Hoch and McCormick 2004). Epidemiological studies conducted in West Africa demonstrated that vaccination with live-attenuated vaccines (LAVs) elicits strong long-term cell-mediated immune responses following single immunization and is the most feasible approach to control LF (Falzarano and Feldmann 2013). It is important to emphasize that vaccination in rural areas of West Africa that have limited infrastructure should provide full protection after a single shot. Live-attenuated vaccines are the best approach that provide the most effective way to present antigens that will induce a broad spectrum of T-cells against all strains of LASV (Lukashevich et al. 2005). Additionally, live-attenuated MOPV/LASV reassortant clone ML29 and vector-based vaccine platforms expressing LASV GPC and/or NP have been shown to induce T-cell-mediated immune responses (Geisbert et al. 2005; Goicochea et al. 2012). Live-attenuated vaccine candidates include Candid#1 vaccine, Mopeia virus (MOPV)/LASV reassortant clone ML29 based vaccine, recombinant Lassa live-attenuated vaccine, and MOPV live-attenuated vaccine (Fig. 3).

4.2.1 Candid#1 Vaccine

Candid#1 is the first live-attenuated vaccine approved for the prevention of arenaviruses. It was developed through serial passaging of pathogenic JUNV isolate XJ in cell culture (Enria and Barrera Oro 2002). The resultant strain was attenuated in mice, guinea pigs, and rhesus macaques resulting in 17 amino acid substitutions distinguishing Candid#1 from the parental XJ13 strain of JUNV (McKee et al. 1993). This vaccine is highly effective and immunogenic (Enria et al. 2010). Candid#1 showed protection from morbidity and mortality from JUNV infection and was approved in Argentina in 2006. Candid#1 is not indicated for pregnant women, immunocompromised individuals, and children less than 15 years of age. A single dose of Candid#1 vaccine administered through intramuscular injection induced JUNV-specific neutralizing antibodies in 91.1% of vaccinated individuals and resulted in 95% vaccine efficacy (Ambrosio et al. 2011). It is unlikely that Candid#1 will be approved in the United States due to the stability of the attenuation phenotype, namely phenylalanine-to-isoleucine substitution at position 427 in the transmembrane of GP2 subunit of the GPC glycoprotein (Gowen et al. 2021a, b).

4.2.2 MOPV/LASV Reassortant Clone ML29 Vaccine

While Mopeia virus (MOPV) is not lethal in humans and was shown to protect NHPs from lethal LASV challenge, LASV, on the other hand, causes mortality in the majority of untreated patients. Reassortants between different arenaviruses were developed to preserve the attenuated profile of MOPV and the ability to induce a robust T-cell-mediated response against different LASV lineages. Clone ML29 encodes a GPC and NP of LASV and the L and the Z protein of MOPV (Lukashevich 1992). ML29 clone was stable in tissue culture and 18 mutations that distinguish ML29 genome from parental LASV (S RNA) and MOPV (L RNA) were preserved (Moshkoff et al. 2007). ML29 was attenuated in guinea pigs and non-human primates. Inoculated strain 13 guinea pigs survived the Josiah strain

LASV challenge administered 30 days post-immunization without the disease and histological lesions. Rhesus macaques (*Macaca mulata*) developed virus-specific T-cells capable of secreting IFN-γ that were detectable in peripheral blood and spleen 7 days after immunization. Cross-reactive ML29-specific cells were detected in 13% of T-cells 28 days post-immunization. Antibody responses were detected by ELISA 2 weeks after vaccination. Neutralizing antibody response was not found. LASV titer was detected on day 5 after challenge in unvaccinated rhesus macaques and high viral load was observed in tissues 15 days post-infection. Although few ML29-vaccinated rhesus macaques had low-titer viremia 5 days post-challenge, none of the tissues of vaccinated rhesus macaques had infectious LASV (Lukashevich et al. 2005). A single-dose immunization with ML29 fully protected strain 13 guinea pigs from lethal challenge with Josiah strain or Nigerian isolate LASV administered 30 days post-immunization. These animals did not display any signs of disease and histological analysis of tissues was normal. Viremia in vaccinated animals was not detectable. Non-neutralizing NP-specific antibodies were detected a week after immunization and peaked after viral challenge (Carrion et al. 2007). Marmosets vaccinated with ML29 and challenged 30 days later with LASV-Josiah were completely protected from infection-inducible morbidity and mortality. ML29 induced low transient viremia in one out of six tested marmosets. LASV was not detectable in blood or tissues in other vaccinated animals after challenge. Vaccinated and challenged marmosets were euthanized 35 days post-challenge without any sign of disease. Unvaccinated marmosets, on the other hand, met euthanasia criteria on days 17–21 post-infection and were sacrificed. Human PBMCs from ML29-treated healthy volunteers were used for gene expression analyses and showed that ML29 is a safe platform for further vaccine development (Lukashevich et al. 2008). LASV is endemic in West Africa, a region with high seroprevalence for infection with HIV-1. Zapata and coworkers infected rhesus macaques with SIV to determine whether ML29 vaccine is safe and immunogenic in rhesus macaques in advanced stages of SIV infection and to model vaccination of HIV-1-infected individuals. Rhesus macaques were infected with SIVmac251 and received ML29 immunizations on days 393 and 429 post-challenge. SIV-infected and uninfected vaccinated macaques did not develop any signs of arenavirus disease. All vaccinated rhesus macaques develop ML29-specific T-cell antibody responses (Zapata et al. 2013).

4.2.3 Recombinant LASV Live-Attenuated Vaccine

Recombinant LASV (rLASV) vaccine candidate ([rLASV (IGR/S-S)] is based on the presence of the noncoding intergenic region (IGR) of the small (S) genome segments (S-IGR) in both large (L) and S LASV segments. Each genomic segment uses an ambisense coding strategy that results in the synthesis of two viral proteins from open reading frames separated by noncoding intergenic regions (IGRs). The small segment encodes NP and GPC (Auperin et al. 1986; Clegg et al. 1991). The large segment encodes the large (L) protein and RING finger protein Z (Djavani et al. 1997; Lukashevich et al. 1997; Perez et al. 2003). In vitro, [rLASV (IGR/S-S)] exhibited moderately lower fitness compared to the wild-type rLASV and was fully attenuated in guinea pigs. A single low-dose subcutaneous administration of [rLASV (IGR/S-S)] provided full protection of strain 13 and Hartley guinea pigs inoculated with a lethal challenge of LASV-WT (Cai et al. 2020).

4.2.4 MOPV Live-Attenuated Vaccine

MOPV is not pathogenic in NHPs and is protective against lethal challenge with LASV (Kiley et al. 1979). Although both LASV and MOPV show tropism for antigen-presenting cells such as dendritic cells and macrophages, MOPV activates antigen-presenting cells, induces type I IFN production and greater T-cell response compared to LASV (Baize et al. 2004, 2006; Pannetier et al. 2004, 2011). Exoribonuclease (ExoN) activity of MOPV NP was shown to be essential for the multiplication of antigen-presenting cells. Mutations introduced in the ExoN site of MOPV NP resulted in the generation of highly attenuated strain $MOPV_{ExoN6b}$. $MOPV_{ExoN6b}$ is more immunogenic than MOPV and induces strong type I IFN responses. Further modifications of this platform that resulted in the expression of heterologous GPC are termed MOPEVAC. One dose of $MOPV_{ExoN6b\ GPC\ LASV}$ ($MOPEVAC_{LAS}$) was used to immunize *C. macaques* followed by Josiah strain LASV challenge 37 days post-immunization. All vaccinated macaques were protected from viral challenge and survived. Some of them were febrile. Unvaccinated animals presented with clinical signs of disease including fever, anorexia, and hypothermia. They were euthanized on days 12 and 15 after viral challenge. TNF-α-producing $CD8^+$ and $CD4^+$ cells were detected in most vaccinated animals 14 days post-immunization. $MOPEVAC_{LAS}$-specific neutralizing antibodies were detected in all immunized animals 23 days post-immunization (Carnec et al. 2018).

4.3 Viral Vector Vaccines

Viral vector vaccines are based on the use of a modified virus as a vector to express a protein that will elicit immune response against the virus of interest. Viral vector vaccines have been developed for Ebola virus, Zika virus, COVID-19, influenza, HIV, and chikungunya virus. Viral vector vaccines elicit good immune responses. One of the disadvantages of viral vector-based vaccines is that in some cases prior immunity to the viral vector may render the vaccine ineffective. Viral vector vaccine candidates against LASV include Vaccinia virus vector vaccines, Vesicular stomatitis virus vector vaccines, Yellow fever 17D virus vector vaccines, Measles virus vector vaccines, Venezuelan Equine Encephalitis Virus vector vaccines,

Rabies virus vector vaccines, and Chimpanzee adenovirus (ChAd) vector vaccines (Fig. 3).

4.3.1 Vaccinia Virus Vector Vaccines

Vaccinia virus-based Lassa vaccines showed inconsistent protective efficacy in guinea pigs and non-human primates (Clegg and Lloyd 1987; Fisher-Hoch et al. 1989, 2000; Morrison et al. 1989). Vaccinia virus (strain Lister) expressing LASV NP (strain GA391) protected Hartley guinea pigs against the LASV challenge. Unvaccinated guinea pigs displayed signs of disease such as anorexia, viremia, pyrexia, and death (Clegg and Lloyd 1987). A recombinant vaccinia virus encoding LASV NP (strain Josiah) protected 94% of strain 13 guinea pigs from homologous LASV challenge. Vaccinia virus expressing LASV GPC protected 79% of guinea pigs infected with LASV. Only 58% of guinea pigs vaccinated with the combination of both vaccines survived the challenge with LASV (Morrison et al. 1989). A recombinant vaccinia virus expressing LASV GP gene (L-LSGPC) protected monkeys from lethal LASV challenge, but the animals experienced transient febrile illness and low-titer viremia (Fisher-Hoch et al. 1989). In a different study, 90% of *C. macaques* that received a recombinant vaccinia virus with the full-length glycoprotein and the nucleoprotein survived. Eighty-eight percent of macaques vaccinated with the full-length glycoprotein survived but did not develop antibodies. All animals that received vaccination with NP developed high titer antibodies, but only 20% survived the lethal LASV challenge. All animals vaccinated with either GP1 or GP2 alone died, but those vaccinated with both GP1 and GP2 survived demonstrating that both glycoproteins are required for protection. Humoral responses did not correlate with vaccine protection from lethal LASV challenge. Although T-cell-mediated immunity was not measured in this study, it was implicated in protecting the animals (Fisher-Hoch et al. 2000; Fisher-Hoch and McCormick 2004). Even though a recombinant vaccinia virus platform showed some promise in animal models, it suppressed the immune response through the inhibition of human dendritic cells. The use of vaccinia virus-based vector for the development of vaccines against LASV endemic in the regions of the world where the population is immuno-suppressed due to HIV-1 positivity poses serious safety concerns.

4.3.2 Vesicular Stomatitis Virus Vector Vaccines

Vesicular stomatitis virus (VSV)-based vaccines are highly immunogenic. Adverse reactions such as fever, headache, chills, myalgia, and fatigue were documented after immunization with VSVΔG/EBOVGP (Heppner et al. 2017; Regules et al. 2017). VSV-based vaccines also require a continuous cold chain for transport and storage, which can be of a disadvantage when used in developing countries.

Replication-competent vaccine based on recombinant vesicular stomatitis virus (rVSV) expressing Lassa viral glycoprotein (VSVΔG/LVGPC) elicited strong humoral and cellular immune responses in *C. macaques* after a single dose administered intramuscularly followed by a lethal Lassa virus (Josiah strain) challenge. Vaccinated *C. macaques* did not show any evidence of clinical disease. Control animals started showing signs of disease on day 3 and succumbed to illness by days 11–13 post-infection. While control animals were viremic until euthanasia, vaccinated *C. macaques* cleared viremia by day 10 (Geisbert et al. 2005). Safronetz and coworkers assessed the ability of rVSV vectors expressing the glycoprotein (rVSVΔG/LASVGPC) and nucleoprotein (rVSVΔG/LASVNP) of LASV Josiah to protect strain 13 guinea pigs and *C. macaques* from morbidity and mortality following LASV isolate Josiah (Sierra Leone, clade IV), Z-132 (Liberia, clade IV), Soromba-R (Mali, clade IV), and Pinneo (Nigeria, clade I) challenge four weeks later. They determined that rVSVΔG/LASVGPC is superior to rVSVΔG/LASVNP as an immunogen and prevented morbidity and mortality in both animal models infected with geographically and genetically distinct Lassa viruses. Guinea pigs vaccinated with rVSVΔG/LASVGPC did not show any signs of illness and were not viremic. While Josiah and Z-132 strains were lethal in all control animals within 16–18 days post-challenge, Soromba-R and Pinneo, on the other hand, were not fully lethal in this animal model. *Cynomolgus macaques* vaccinated with rVSVΔG/LASVGPC and challenged with Z-132 did not display any signs of disease and the virus was not detected compared to control animals that displayed the signs of illness 5–7 days post-infection and progressed to terminal illness by day 13 (Safronetz et al. 2015a, b; Warner et al. 2018). Stein and coworkers assessed the ability of rVSVΔG/LASVGPC to induce long-term immunity in a guinea pig-adapted (GPA) LASV model of infection. In this study, outbred Hartley guinea pigs were immunized with a single dose of rVSVΔG/LASVGPC followed by a lethal challenge of GPA-LASV at 7, 14, 25, 189, and 355 days post-vaccination. While full protection was achieved at 7- and 14 days post-infection, 83%, 87%, and 71% protection were achieved at 25 days, 189 days, and 355 days post-infection, respectively. Total LASVGPC-specific IgG antibodies were measured every month after vaccination. Antibody titers peaked by day 51 and sustained until sampling days 171 and 351 post-challenge. Vaccinated animals had higher antibody titers on day 21 post-infection compared to the animals that succumbed to disease. Vaccinated guinea pigs had significantly lower levels of infectious virus in serum, lung, and spleen on days 7 and 14 post-challenge compared to control animals on days 10 and 12 post-infection. This study demonstrated that a single dose of rVSVΔG/LASVGPC can induce long-term protection from LASV infection (Stein et al. 2019).

VesiculoVax is a quadrivalent formulation consisting of three live, attenuated, replication-competent rVSV vectors that encode EBOV (Kikwit strain), SUDV (Boniface strain), and MARV (Angola strain) glycoproteins and live, attenuated, replication-competent rVSV vector expressing LASV (Josiah strain) surface glycoprotein. VSV vector contains a translocated N gene and a truncated cytoplasmic tail. This vaccine platform was developed by Profectus BioSciences to improve the safety profile of the VSV vector. VesiculoVax vector was shown to cause the expansion of B-cells to antibody-producing cells. VesiculoVax vector was used to generate vaccines against Ebola and Marburg virus (Mire et al. 2015; Matassov et al. 2018; Cross et al. 2020; Woolsey et al. 2022). Two doses of VesiculoVax protected *C. macaques* from a lethal challenge with a heterologous strain of LASV (lineage II). Glycoprotein-specific neutralizing antibodies and IgG responses were detected in serum after the vaccination. Additionally, cell-mediated immune responses specific to glycoproteins were also detected (Cross et al. 2020).

4.3.3 Yellow Fever 17D Virus Vector Vaccine

Yellow fever 17D (YF17D) vector has been used as a platform for Dengue virus, West Nile virus, and Japanese encephalitis vaccines. LASV GPC (strain AV) was cloned between YF17D E and NS1 genes. The YF17D/LASVΔGPC recombinant virus was replication-competent, attenuated, and elicited immune responses against YF and LASV. A single subcutaneous administration of YF17D/LASVΔGPC protected 80% of strain 13 guinea pigs from lethal Josiah strain LASV (Bredenbeek et al. 2006). However, duplication of flavivirus sequences flanking the inserted gene was contributing to the instability of YF17D/LASVΔGPC. Size reduction of ΔGPC insert and cloning of GP1 and GP2 between YF17D E and NS1 gene into separate vectors resulted in genetic stability of recombinant viruses. Vaccination of strain 13 guinea pigs with YF17D/LASV-GP1 and YF17D/LASV-GP2 protected 83% of animals against fatal LF (Jiang et al. 2011). Two-dose vaccination of marmosets with YF17D/LASVΔGPC did not protect the animals from a lethal challenge with Josiah LASV (Lukashevich 2012). Reformulation of this vaccine to improve its immunogenicity is needed before it is considered as a justifiable option to control Lassa fever and Yellow fever in West Africa.

4.3.4 Measles Virus Vector Vaccines

Live-attenuated measles virus (MV) vaccine is the most effective and safe. It induces both humoral and cellular responses and provides long-lasting immunity after a prime-boost vaccination regimen with the efficacy of 97%. MV is produced on a large scale and is distributed at a low cost in many countries. The accessibility of the MV production provides the means for rapid stockpiling of the vaccine during epidemic. Attenuated MV strains such as the Schwartz/Moraten, Edmonston-Zagreb, and AIK-C, are great platforms for the development of recombinant MV vaccine candidates against viral infections (Frantz et al. 2018). MV vector was used for the development of vaccines against Chikungunya virus, Dengue virus, Zika virus, SARS-CoV, and Nipah virus (Frantz et al. 2018).

Mateo and coworkers developed the Schwartz strain MV vector-based vaccines expressing GPC alone or in combination with NP or Z and accessed the ability of a single administration of these vaccines to protect *C. macaques* from Josiah strain

LASV. Monkeys were challenged 37 days after immunization. The vaccination was well tolerated. Viral RNA was not detected in either plasma, peripheral blood, nasal, or oral secretions. The most effective vaccine candidate was the MV vector expressing LASV glycoprotein and nucleoprotein (MeV-NP). Complete protection from viral challenge was associated with T-cell and antibody responses against LASV. GPC-specific $CD8^+$ and $CD4^+$ T-cells were detected 23 days after immunization with MeV-NP. An increase in NP-specific T-cells was noted 14 days after immunization with MeV-NP. Low titers of LASV-specific IgG were detected in 50% of *C. macaques* immunized with MeV-NP 37 days after immunization (Mateo et al. 2019). In a follow-up study, MeV-NP vaccine was evaluated for efficacy against the challenge of *C. macaques* with heterologous strains of LASV from lineage II Nigerian strain or lineage VII Benin strain of LASV 37 days post-vaccination. Monkeys were immunized through the subcutaneous route. The route of administration did not have any effect on the biological parameters between vaccination and challenge. MeV-NP that was developed using Josiah strain of LASV was protective against Nigerian and Benin strains. An additional group of MeV-NP-vaccinated monkeys that were immunized through either subcutaneous or intramuscular route was challenged 1 year later with the Josiah strain LASV. A single immunization dose of MeV-NP was sufficient for the generation of a long-lasting protective immune response against LASV. This vaccine candidate progressed to evaluation in phase I clinical trials (Mateo et al. 2021). It has been previously noted that if LASV vaccine platform that expresses LASV NP is protective and is approved for use in LASV endemic area, it might present a problem in distinguishing infected individuals from vaccinated ones since the Lassa diagnostic tests are mostly based on NP-specific responses (Meulen et al. 2004).

4.3.5 Venezuelan Equine Encephalitis Virus Vector Vaccines

Venezuelan Equine Encephalitis Virus (family *Togaviridae*; Alphavirus genus) is a positive-sense RNA virus. Like other alphaviruses including Semliki virus and Sindbis virus, Venezuelan Equine Encephalitis Virus (VEEV) has been successfully used in the development of vaccine platforms (Xiong et al. 1989; Liljeström and Garoff 1991). Alphavirus replicon particles (RP) are one of the expression systems that is widely used for vaccine development. They are single-cycle replication-deficient vectors that are generated by the deletion of structural genes from the replicon RNA and their replacement with the heterologous genes of interest. Replicon RNA and helper genes in *trans* are transfected into permissive cells resulting in the packaging of virus-like particles (VLPs) expressing encoded antigens. Alphavirus RP has been studied in different animal models and demonstrated good safety and immunogenicity profile (Vander Veen et al. 2012). VEEV RP demonstrated potent adjuvant activity for systemic and mucosal antibody responses and elicited mucosal antigen-specific $CD8^+$ T-cell responses (Thompson et al. 2008). VEEV RP expressing LASV-NP, LASV-GPC, and LASV-GPC+NP protected strain 13 guinea pigs from Josiah strain LASV challenge. Low-level viremia was detected on day 7 after the challenge in a few animals in each group. Immunization with LASV-GPC+NP elicited a robust antibody response (Pushko et al. 2001). VEEV RP was also used to express a bivalent vaccine expressing LASV and EBOV glycoprotein genes. Immunization of guinea pigs with VEEV RP expressing glycoprotein of LASV and EBOV protected the animals against challenges with both viruses (Pushko et al. 1997, 2001).

To improve VEEV RP vaccine safety and cloning capacity, human vaccine VEEV TC-83 backbone was used to generate replication restricted bicistronic virus-like particle vectors expressing GPCs of both JUNV and MACV and was designated JV&MV VLPV vaccine. Hartley guinea pigs received two doses of candidate vaccine 28 days apart. Vaccinated guinea pigs were protected from the Strain Romero JUNV challenge. JV&MV VLPV-immunized guinea pigs had normal to limited histological changes and very low-level JUNV was only detectible in the spleen tissues. Mock-vaccinated animals, on the other hand, developed fever on day 7 and weight loss on day 8 after the JUNV challenge and became lethargic, and some of them developed hind-limb paralysis on day 11 post-infection. JV&MV VLPV-vaccinated and challenged with Carvallo strain MACV guinea pigs did not display any signs of disease and gained weight. Minimal histological changes were observed. Mock-vaccinated guinea pigs lost appetite, started losing weight, and became febrile 8–10 days post-infection. This study demonstrated that JV&MV VLPV is a good and effective alternative to Candid#1 vaccine. This vaccine candidate is a replication-restricted vehicle that allows the expression of antigens of interest in targeted cells such as dendritic cells and macrophages, which provides an immunological advantage. As described by Wang and coworkers, VEEV VLPV expressing LASV GPCwt and ΔGPfib were immunogenic in mice and upregulated $CD11c^+/CD8^+$ dendritic cells that play an important role in cross-presentation (Johnson et al. 2020).

4.3.6 Rabies Virus Vector Vaccines

Rabies virus (RABV) (family *Rhabdoviridae*) is a negative-sense, single-stranded RNA virus. RABV causes up to 55,000 human deaths annually, mostly in Asia and Africa. The most affected by this disease are children (Chidumayo 2018). Rabies vaccine is very effective and is administered as preexposure and postexposure prophylactic. However, rabies vaccine is costly to produce and must be given several times, which can be difficult to achieve in remote areas of Africa. An ideal vaccine in this situation is a temperature-stable vaccine that can elicit a long-lasting protective immune response after a single immunization.

The attenuated RABV vaccine strain SAD-B19 was further attenuated by the introduction of a mutation within RBV GP followed by chemical inactivation (BNSP333) (McGettigan et al. 2003). Codon-optimized LASV-GPC was cloned into BNSP333 between RABV N and P genes and was designated as LASSARAB. Lassa virus and rabies vaccine (LASSARAB) that expressed a codon-optimized LASV glycoprotein (coGPC) and rabies virus G was adjuvanted by a TLR-4 agonist (GLA-SE). LASSARAB protected both mice and guinea pigs against Lassa fever and elicits humoral response against LASV and RABV in both animal models. LASSARAB elicited high LASV GPC-specific IgG antibodies that correlate with protection before LASV challenge in the absence of neutralizing antibodies. This study demonstrated that non-neutralizing antibodies against LASV GPC play an important role in protection against LASV challenge in a mouse model (Abreu-Mota et al. 2018). The same group of researchers tested a tetravalent RABV-vectored vaccine candidate against EBOV, MARV, SUDV, and LASV GPs (FILORAB1-3 and LASSARAB). NHPs were also immunized with monovalent RABV-vectored vaccine candidates against each of the viruses. *Cynomolgus macaques* were immunized and adjuvanted twice 28 days apart. The humoral responses were followed for one year. Animals vaccinated with the tetravalent vaccine elicited LASV-GPC IgG titers that peaked on day 56 and were not significantly different from the LASV-GPC IgG titers detected in animals immunized with LASSARAB only. LASSARAB alone and FILORAB1-3 and LASSARAB administered together elicited a long-lasting LASV-GPC-specific response lasting for up to 365 days. Importantly, FILORAB1 (RABV-vectored Ebola virus) was observed to be stable at 4 and 37 °C for 6 months and 50 °C for 2 weeks. LASSARAB vaccine candidate is prepared using the same protocol and should be stable similarly to FILORAB1 (Kurup et al. 2021).

4.3.7 Chimpanzee Adenovirus Vector Vaccines

Recombinant adenoviruses have been used as vaccine platforms for various pathogens. Simian adenoviral vectors are an attractive option for the development of vaccines because of the absence of vector immunity in the human population. The Chimpanzee adenovirus vector ChAdOx1 is derived from chimpanzee adenovirus Y25. Deletion of the transcriptional unit E1 results in replication incompetent phenotype of the vector and deletion of the E3 unit allows the insertion of heterologous genes of interest up to 8 kb in size (Purushotham et al. 2019). ChAdOx1-based vaccines are efficacious against Zika virus, Rift Valley Virus, Middle East respiratory syndrome coronavirus, and Mycobacterium tuberculosis (Warimwe et al. 2016; Munster et al. 2017; López-Camacho et al. 2018; Stylianou et al. 2018). ChAdOx1 exhibits a good safety profile as demonstrated in clinical trials in subjects of 5 months to over 70 years of age. Additionally, ChAdOx1 is thermostable up to 45 °C, which is an important characteristic of the vaccine platform that could be used to control pathogens in areas with unstable cold-chain access.

ChAdOx1-Lassa-GPC is a chimpanzee adenovirus vector vaccine expressing LASV GPC (Josiah strain). ChAdOx1-Lassa-GPC induced polyfunctional T-cell responses in CD-1 mice. Vaccine-induced T-cell and antibody responses exhibited cross-reactivity to GP from diverse LASV strains including Pinneo (lineage I), 803,213 (lineage II), and GA391 (lineage III). Hartley guinea pigs immunized with a single dose of ChAdOx1-Lassa-GPC protected animals against morbidity and mortality after infection with a guinea pig-adopted Josiah strain LASV compared to mock-vaccinated animals. GP-specific IgG antibodies were detected 28 days post-immunization in guinea pigs. An increase in IgG antibody titer was observed after the administration of the second dose of ChAdOx1-Lassa-GPC. Vaccination with ChAdOx1-Lassa-GPC did not induce a significant neutralizing antibody response. Limited pathology was observed in hematoxylin and eosin-stained lung, liver, and spleen tissues collected from vaccinated animals on day 12 after challenge. LASV-specific in situ hybridization (ISH) did not detect replicating virus. Lung sample from one guinea pig vaccinated twice showed low level of ISH staining for positive-strand LASV RNA. The results of this study demonstrate that ChAdOx1-Lassa-GPC is a good candidate to investigate the efficacy of this candidate vaccine against other strains of LASV (Fischer et al. 2021).

4.4 DNA Vaccines

DNA vaccines are stable and can be produced quickly in response to the outbreak. This approach is based on the cloning of the gene that encodes the antigen of interest into the plasmid. LASV GPC gene (Josiah strain) was codon-optimized to enhance the expression in the recipient species and cloned into pWRG7077 expression vector (pLASV-GPC). The vaccine was delivered three times by dermal electroporation in a guinea pig model (strain 13) at 3-week intervals. Guinea pigs were protected from viremia, illness, and death after a lethal dose of Josiah strain LASV. Unvaccinated and mock-vaccinated guinea pigs developed illness and succumbed by day 18 post-infection. Neutralizing antibodies were detected in vaccinated animals that peaked 3 weeks post-challenge. A small group of vaccinated guinea pigs were challenged for the second time with a lethal dose of Josiah strain LASV 120 days after the endpoint of the study and were monitored for the disease symptoms for 30 days post-challenge. Vaccinated animals did not develop any signs of disease in comparison to unvaccinated guinea pigs that showed signs of febrile illness and succumbed to disease by days 15–18 post-infection (Cashman et al. 2013). The efficacy of pLASV-GPC vaccine was investigated in *C. macaques* comparing a three-dose schedule to a two-dose schedule pLASV-GPC vaccine delivered by dermal electroporation 3 or 4 weeks apart. Monkeys were challenged with LASV Josiah strain five weeks after the last vaccine administration. Animals were monitored for 45 days for signs of disease and viremia. All vaccinated NHPs survived the challenge, did not display disease, and were not viremic. Thirty percent of mock-vaccinated animals that survived the challenge became chronically ill. The subset of those that survived experienced sensorineural hearing loss. This vaccine candidate progressed to evaluation in phase I clinical trials (Cashman et al. 2017). LASV GPC-specific neutralizing antibodies and T-cell responses were detected in NHPs after two immunizations with pLASV-GPC spaced 4 weeks apart (Jiang et al. 2019) (Fig. 3).

4.5 Virus-Like Particles (VLP) Vaccines

Virus-like particle (VLP)-based platforms have been successfully used for the development of vaccines against several viruses. Gardasil, a recombinant quadrivalent VLP-based vaccine against human Papillomavirus (types 6, 11, 16, and 18), was developed by Merck & Co., Inc. Gardasil is immunogenic and well tolerated in humans (Medeiros et al. 2009; Block et al. 2010). ENGERIX-B is a recombinant VLP-like Hepatitis B vaccine developed by GlaxoSmithKline plc. ENGERIX-B is very efficacious, highly immunogenic, and safe (Elhanan et al. 2018). VLP-based vaccines have also been developed and tested to prevent other viruses such as EBOV (Swenson et al. 2005), Influenza (Pushko et al. 2007, 2010) (Mahmood et al. 2008), MARV (Swenson et al. 2005; Warfield et al. 2007), Dengue virus (Murawski et al. 2010), West Nile virus (Spohn et al. 2010), Hepatitis C (Mihailova et al. 2006), HIV (Kang et al. 1999), and Chikungunya (Akahata et al. 2010).

Branco and coworkers generated LASV VLP expressing LASV GPC, NP, and Z matrix protein with post-translational modifications using a mammalian expression system. BALB/C mice were immunized with VLP Z+GPC+NP or VLP Z+GPC through intraperitoneal injection in a prime plus 2 boost schedule 3 weeks apart without an adjuvant. Lassa VLP Z+GPC+NP induced significantly higher IgG antibody response to the individual viral proteins compared to VLP Z+GPC. Additionally, LASV VLP was also evaluated as a diagnostic tool for the detection of viral protein-specific IgM and IgG in the serum of convalescent Lassa patients using ELISA. Sera from convalescent Lassa fever patients VLP antigens were used to confirm that the proteins on the surface of VLP were displayed in the native conformation (Branco et al. 2010). VLP generated from a Modified Vaccinia Ankara Vector expressing LASV GPC and Z from strain Josiah (GEO-LM01) fully protected CBA/J mice from a lethal dose of ML29 reassortant virus that was delivered intracranially 14 days post-immunization. Mock-vaccinated mice succumbed to disease 7 days after challenge. GEO-LM01 vaccine candidate elicited a strong T-cell response and low levels of antibodies (Salvato et al. 2019). LASV GP VLPs derived from MDCK II cell line stably expressing LASV GPC strain Josiah administered with Sigma Adjuvant system elicited neutralizing antibody response in New Zealand White rabbits in a prime plus 3 boost schedule (4, 7, and 10 weeks later) (Müller et al. 2020). Polyclonal neutralizing antibodies displayed cross-reactivity against five genetic lineages of LASV, namely lineages I, II, V, VII, and LCMV. These antibodies also blocked virus infection at the attachment and postattachment steps in different human cell lines infected with VSVΔG/LASVGP (Müller et al. 2020) (Fig. 3).

4.6 *Epitope-Based Vaccines*

Heterogeneity of arenaviruses presents a significant challenge in the development of an effective vaccine. LASV GPC and NP have been commonly used as immunogens for the development of candidate vaccine platforms (Bowen et al. 2000). However, most vaccines that are in preclinical development right now express only GPC from Josiah strain. Recent studies have demonstrated that immune protection during LASV infection is likely conferred by cell-mediated response dependent on Type I IFN (Meulen et al. 2004; Botten et al. 2006).

HLA-A2 epitopes of LASV GPC peptides LASV GPC_{42-50}, LASV GPC_{60-68}, and LASV $GPC_{441-449}$ were identified by utilizing a predictive algorithm-based approach induced potent $CD8^+$ T-cell response in HLA-A*0201 transgenic mice. HLA-A*0201 transgenic mice immunized with LASV GPC_{42-50} and LASV GPC_{60-68} were protected from challenge with recombinant vaccinia virus expressing LASV GPC 14 days post-immunization (Botten et al. 2006). Kotturi and coworkers attempted to design a universal vaccine capable of eliciting HLA-restricted $CD8^+$ T-cell responses against seven pathogenic arenaviruses, namely LCMV, Lassa, Guanarito, Junin, Machupo, Sabia, and Whitewater Arroyo viruses. HLA-A*0201 transgenic mice were inoculated with rVACV expressing GPC, NP, Z, or L from 7 different arenaviruses. Twelve epitopes from 6 arenaviruses were identified: LASV $GPC_{111-120}$, JUNV GPC_{7-15}, JUNV GPC_{18-26}, LCMV GPC_{34-43}, MACV GPC_{18-26}, SABV $GPC_{142-150}$, WWAV GPC_{42-50}, MACV NP_{19-27}, MACV $NP_{432-440}$, SABV $NP_{547-556}$, WWAV $NP_{274v282}$, and LCMV Z_{49-58}. Immunization of HLA-A*0201 transgenic mice with these epitopes resulted in robust $CD8^+$ T-cell responses for 7 of the 12 peptides (LASV $GPC_{111-120}$, SABV $GPC_{142-150}$, WWAV GPC_{42-50}, MACV NP_{19-27}, SABV $NP_{547-556}$, WWAV $NP_{274-282}$, and LCMV Z_{49-58}) and protected mice from rVACV challenge expressing Old and New World arenavirus glycoproteins (Kotturi et al. 2009). In another study, HLA class I-restricted cross-reactive epitopes LASV $GPC_{441-449}$ and LCMV $GPC_{447-455}$ were identified and could be used for the development of vaccine candidates against pathogenic arenaviruses (Botten et al. 2010) (Fig. 3).

5 Conclusions

Significant efforts have been made in the development of LASV diagnostics. Timely diagnosis of LF requires a combination of a rapid point-of-care, serological, and molecular diagnostic tools. Attempts to improve efficacy and increase sensitivity of already available diagnostics tools are necessary to detect different strains of LASV and distinguish LF from other diseases with similar clinical presentation.

Although over 50 years have passed since the discovery of LASV, LF therapeutics are currently in preclinical development. Human monoclonal antibody cocktail Arevirumab exhibited exceptional efficacy in guinea pigs and NHPs and should be considered for clinical trials. However, the production of human monoclonal antibodies remains to be high and cost-prohibitive for LASV endemic areas. Favipiravir showed protection in mice, guinea pigs, and NHPs and its efficacy should be evaluated in clinical trials. Small molecules like favipiravir that interferes with viral genome replication generally exhibit broader efficacy compared to monoclonal antibodies that are more strain specific.

Vaccines are the most effective ways to control viral infections. Development of the vaccine against LASV has been an enormous endeavor for the past 50 years. Most of the vaccine candidates are in preclinical development and demonstrated efficacy in animal models. Only recently two vaccine candidates, MeV-NP and pLASV-GPC, entered the evaluation in phase I clinical trials. One of the main challenges of LASV vaccine development is the ability of a candidate vaccine to elicit both broadly neutralizing antibodies and T-cell-mediated immune responses. Vaccine vector expressing LASV GPC in prefusion conformation will elicit the response of broadly neutralizing antibodies. LASV NP has been shown to elicit T-cell responses. More studies should be done on the development of a vaccine that expresses both GPC and NP to achieve a high level of protection against different strains of LASV. Additionally, establishment of infrastructure and training of personnel for vaccine production in LASV endemic areas is of great importance.

References

Abreu-Mota T, Hagen KR, Cooper K, Jahrling PB, Tan G, Wirblich C, Johnson RF, Schnell MJ (2018) Non-neutralizing antibodies elicited by recombinant Lassa-Rabies vaccine are critical for protection against Lassa fever. Nat Commun 9(1):4223

Akahata W, Yang ZY, Andersen H, Sun S, Holdaway HA, Kong WP, Lewis MG, Higgs S, Rossmann MG, Rao S, Nabel GJ (2010) A virus-like particle vaccine for epidemic Chikungunya virus protects nonhuman primates against infection. Nat Med 16(3):334–338

Amanat F, Duehr J, Huang C, Paessler S, Tan GS, Krammer F (2020) Monoclonal antibodies with neutralizing activity and Fc-effector functions against the machupo virus glycoprotein. J Virol 94(5)

Ambrosio A, Saavedra M, Mariani M, Gamboa G, Maiza A (2011) Argentine hemorrhagic fever vaccines. Hum Vaccin 7(6):694–700

Ambrosio AM, Enria DA, Maiztegui JI (1986) Junin virus isolation from lympho-mononuclear cells of patients with Argentine hemorrhagic fever. Intervirology 25(2):97–102

Auperin DD, Sasso DR, McCormick JB (1986) Nucleotide sequence of the glycoprotein gene and intergenic region of the Lassa virus S genome RNA. Virology 154(1):155–167

Baize S, Kaplon J, Faure C, Pannetier D, Georges-Courbot MC, Deubel V (2004) Lassa virus infection of human dendritic cells and macrophages is productive but fails to activate cells. J Immunol 172(5):2861–2869

Baize S, Pannetier D, Faure C, Marianneau P, Marendat I, Georges-Courbot MC, Deubel V (2006) Role of interferons in the control of Lassa virus replication in human dendritic cells and macrophages. Microbes Infect 8(5):1194–1202

Bausch DG, Rollin PE, Demby AH, Coulibaly M, Kanu J, Conteh AS, Wagoner KD, McMullan LK, Bowen MD, Peters CJ, Ksiazek TG (2000) Diagnosis and clinical virology of Lassa fever as evaluated by enzyme-linked immunosorbent assay, indirect fluorescent-antibody test, and virus isolation. J Clin Microbiol 38(7):2670–2677

Blaising J, Polyak SJ, Pécheur EI (2014) Arbidol as a broad-spectrum antiviral: an update. Antiviral Res 107:84–94

Block SL, Brown DR, Chatterjee A, Gold MA, Sings HL, Meibohm A, Dana A, Haupt RM, Barr E, Tamms GM, Zhou H, Reisinger KS (2010) Clinical trial and post-licensure safety profile of a prophylactic human papillomavirus (types 6, 11, 16, and 18) l1 virus-like particle vaccine. Pediatr Infect Dis J 29(2):95–101

Boisen ML, Hartnett JN, Shaffer JG, Goba A, Momoh M, Sandi JD, Fullah M, Nelson DKS, Bush DJ, Rowland MM, Heinrich ML, Koval AP, Cross RW, Barnes KG, Lachenauer AE, Lin AE, Nekoui M, Kotliar D, Winnicki SM, Siddle KJ, Gbakie M, Fonnie M, Koroma VJ, Kanneh L, Kulakosky PC, Hastie KM, Wilson RB, Andersen KG, Folarin OO, Happi CT, Sabeti PC, Geisbert TW, Saphire EO, Khan SH, Grant DS, Schieffelin JS, Branco LM, Garry RF (2018) Field validation of recombinant antigen immunoassays for diagnosis of Lassa fever. Sci Rep 8(1):5939

Borenstein-Katz A, Shulman A, Hamawi H, Leitner O, Diskin R (2019) Differential antibody-based immune response against isolated GP1 receptor-binding domains from Lassa and Junín Viruses. J Virol 93(8)

Botten J, Alexander J, Pasquetto V, Sidney J, Barrowman P, Ting J, Peters B, Southwood S, Stewart B, Rodriguez-Carreno MP, Mothe B, Whitton JL, Sette A, Buchmeier MJ (2006) Identification of protective Lassa virus epitopes that are restricted by HLA-A2. J Virol 80 (17):8351–8361

Botten J, Whitton JL, Barrowman P, Sidney J, Whitmire JK, Alexander J, Kotturi MF, Sette A, Buchmeier MJ (2010) A multivalent vaccination strategy for the prevention of Old World arenavirus infection in humans. J Virol 84(19):9947–9956

Bowen MD, Rollin PE, Ksiazek TG, Hustad HL, Bausch DG, Demby AH, Bajani MD, Peters CJ, Nichol ST (2000) Genetic diversity among Lassa virus strains. J Virol 74(15):6992–7004

Branco LM, Grove JN, Boisen ML, Shaffer JG, Goba A, Fullah M, Momoh M, Grant DS, Garry RF (2011) Emerging trends in Lassa fever: redefining the role of immunoglobulin M and inflammation in diagnosing acute infection. Virol J 8:478

Branco LM, Grove JN, Geske FJ, Boisen ML, Muncy IJ, Magliato SA, Henderson LA, Schoepp RJ, Cashman KA, Hensley LE, Garry RF (2010) Lassa virus-like particles displaying all major immunological determinants as a vaccine candidate for Lassa hemorrhagic fever. Virol J 7:279

Bredenbeek PJ, Molenkamp R, Spaan WJ, Deubel V, Marianneau P, Salvato MS, Moshkoff D, Zapata J, Tikhonov I, Patterson J, Carrion R, Ticer A, Brasky K, Lukashevich IS (2006) A recombinant Yellow Fever 17D vaccine expressing Lassa virus glycoproteins. Virology 345 (2):299–304

Brouillette RB, Phillips EK, Ayithan N, Maury W (2017) Differences in Glycoprotein complex receptor binding site accessibility prompt poor cross-reactivity of neutralizing antibodies between closely related arenaviruses. J Virol 91(7)

Cai Y, Ye C, Cheng B, Nogales A, Iwasaki M, Yu S, Cooper K, Liu DX, Hart R, Adams R, Brady T, Postnikova EN, Kurtz J, St Claire M, Kuhn JH, de la Torre JC, Martínez-Sobrido L (2020) A Lassa fever live-attenuated vaccine based on codon deoptimization of the viral glycoprotein gene. mBio 11(1)

Carnec X, Mateo M, Page A, Reynard S, Hortion J, Picard C, Yekwa E, Barrot L, Barron S, Vallve A, Raoul H, Carbonnelle C, Ferron F, Baize S (2018) A vaccine platform against arenaviruses based on a recombinant hyperattenuated mopeia virus expressing heterologous glycoproteins. J Virol 92(12)

Carrion R Jr, Patterson JL, Johnson C, Gonzales M, Moreira CR, Ticer A, Brasky K, Hubbard GB, Moshkoff D, Zapata J, Salvato MS, Lukashevich IS (2007) A ML29 reassortant virus protects guinea pigs against a distantly related Nigerian strain of Lassa virus and can provide sterilizing immunity. Vaccine 25(20):4093–4102

Cashman KA, Broderick KE, Wilkinson ER, Shaia CI, Bell TM, Shurtleff AC, Spik KW, Badger CV, Guttieri MC, Sardesai NY, Schmaljohn CS (2013) Enhanced efficacy of a

codon-optimized DNA vaccine encoding the glycoprotein precursor gene of Lassa virus in a guinea pig disease model when delivered by dermal electroporation. Vaccines 1(3):262–277
Cashman KA, Smith MA, Twenhafel NA, Larson RA, Jones KF, Allen RD 3rd, Dai D, Chinsangaram J, Bolken TC, Hruby DE, Amberg SM, Hensley LE, Guttieri MC (2011) Evaluation of Lassa antiviral compound ST-193 in a guinea pig model. Antiviral Res 90 (1):70–79
Cashman KA, Wilkinson ER, Shaia CI, Facemire PR, Bell TM, Bearss JJ, Shamblin JD, Wollen SE, Broderick KE, Sardesai NY, Schmaljohn CS (2017) A DNA vaccine delivered by dermal electroporation fully protects cynomolgus macaques against Lassa fever. Hum Vaccin Immunother 13(12):2902–2911
Chidumayo NN (2018) System dynamics modelling approach to explore the effect of dog demography on rabies vaccination coverage in Africa. PLoS ONE 13(10):e0205884
Chou YY, Cuevas C, Carocci M, Stubbs SH, Ma M, Cureton DK, Chao L, Evesson F, He K, Yang PL, Whelan SP, Ross SR, Kirchhausen T, Gaudin R (2016) Identification and characterization of a novel broad-spectrum virus entry inhibitor. J Virol 90(9):4494–4510
Clark LE, Mahmutovic S, Raymond DD, Dilanyan T, Koma T, Manning JT, Shankar S, Levis SC, Briggiler AM, Enria DA, Wucherpfennig KW, Paessler S, Abraham J (2018) Vaccine-elicited receptor-binding site antibodies neutralize two new world hemorrhagic fever arenaviruses. Nat Commun 9(1):1884
Clegg JC, Lloyd G (1987) Vaccinia recombinant expressing Lassa-virus internal nucleocapsid protein protects guineapigs against Lassa fever. Lancet 2(8552):186–188
Clegg JC, Wilson SM, Oram JD (1991) Nucleotide sequence of the S RNA of Lassa virus (Nigerian strain) and comparative analysis of arenavirus gene products. Virus Res 18(2–3):151–164
Cohen-Dvashi H, Amon R, Agans KN, Cross RW, Borenstein-Katz A, Mateo M, Baize S, Padler-Karavani V, Geisbert TW, Diskin R (2020) Rational design of universal immunotherapy for TfR1-tropic arenaviruses. Nat Commun 11(1):67
Cresta B, Padula P, de Martinez Segovia M (1980) Biological properties of Junin virus proteins. I. Identification of the immunogenic glycoprotein. Intervirology 13(5):284–288
Cross RW, Hastie KM, Mire CE, Robinson JE, Geisbert TW, Branco LM, Ollmann Saphire E, Garry RF (2019) Antibody therapy for Lassa fever. Curr Opin Virol 37:97–104
Cross RW, Mire CE, Branco LM, Geisbert JB, Rowland MM, Heinrich ML, Goba A, Momoh M, Grant DS, Fullah M, Khan SH, Robinson JE, Geisbert TW, Garry RF (2016) Treatment of Lassa virus infection in outbred guinea pigs with first-in-class human monoclonal antibodies. Antiviral Res 133:218–222
Cross RW, Xu R, Matassov D, Hamm S, Latham TE, Gerardi CS, Nowak RM, Geisbert JB, Ota-Setlik A, Agans KN, Luckay A, Witko SE, Soukieh L, Deer DJ, Mire CE, Feldmann H, Happi C, Fenton KA, Eldridge JH, Geisbert TW (2020) Quadrivalent VesiculoVax vaccine protects nonhuman primates from viral-induced hemorrhagic fever and death. J Clin Invest 130 (1):539–551
Cummins D, Bennett D, Machin SJ (1991) Exchange transfusion of a patient with fulminant Lassa fever. Postgrad Med J 67(784):193–194
Delang L, Abdelnabi R, Neyts J (2018) Favipiravir as a potential countermeasure against neglected and emerging RNA viruses. Antiviral Res 153:85–94
Demby AH, Chamberlain J, Brown DW, Clegg CS (1994) Early diagnosis of Lassa fever by reverse transcription-PCR. J Clin Microbiol 32(12):2898–2903
Djavani M, Lukashevich IS, Sanchez A, Nichol ST, Salvato MS (1997) Completion of the Lassa fever virus sequence and identification of a RING finger open reading frame at the L RNA 5' end. Virology 235(2):414–418
Drosten C, Kümmerer BM, Schmitz H, Günther S (2003) Molecular diagnostics of viral hemorrhagic fevers. Antiviral Res 57(1–2):61–87
Eberhardt KA, Mischlinger J, Jordan S, Groger M, Günther S, Ramharter M (2019) Ribavirin for the treatment of Lassa fever: a systematic review and meta-analysis. Int J Infect Dis 87:15–20

Elhanan E, Boaz M, Schwartz I, Schwartz D, Chernin G, Soetendorp H, Gal Oz A, Agbaria A, Weinstein T (2018) A randomized, controlled clinical trial to evaluate the immunogenicity of a PreS/S hepatitis B vaccine Sci-B-Vac™, as compared to Engerix B(®), among vaccine naïve and vaccine non-responder dialysis patients. Clin Exp Nephrol 22(1):151–158

Enria DA, Ambrosio AM, Briggiler AM, Feuillade MR, Crivelli E (2010) Candid#1 vaccine against argentine hemorrhagic fever produced in Argentina. Immunogenicity and safety. Medicina (b Aires) 70(3):215–222

Enria DA, Barrera Oro JG (2002) Junin virus vaccines. Curr Top Microbiol Immunol 263:239–261

Enria DA, Briggiler AM, Fernandez NJ, Levis SC, Maiztegui JI (1984) Importance of dose of neutralising antibodies in treatment of Argentine haemorrhagic fever with immune plasma. Lancet 2(8397):255–256

Enria DA, Briggiler AM, Sánchez Z (2008) Treatment of Argentine hemorrhagic fever. Antiviral Res 78(1):132–139

Enria DA, Maiztegui JI (1994) Antiviral treatment of Argentine hemorrhagic fever. Antiviral Res 23(1):23–31

Falzarano D, Feldmann H (2013) Vaccines for viral hemorrhagic fevers-progress and shortcomings. Curr Opin Virol 3(3):343–351

Feld JJ, Hoofnagle JH (2005) Mechanism of action of interferon and ribavirin in treatment of hepatitis C. Nature 436(7053):967–972

Fischer RJ, Purushotham JN, van Doremalen N, Sebastian S, Meade-White K, Cordova K, Letko M, Jeremiah Matson M, Feldmann F, Haddock E, LaCasse R, Saturday G, Lambe T, Gilbert SC, Munster VJ (2021) ChAdOx1-vectored Lassa fever vaccine elicits a robust cellular and humoral immune response and protects guinea pigs against lethal Lassa virus challenge. NPJ Vaccines 6(1):32

Fisher-Hoch SP, Hutwagner L, Brown B, McCormick JB (2000) Effective vaccine for Lassa fever. J Virol 74(15):6777–6783

Fisher-Hoch SP, McCormick JB (2004) Lassa fever vaccine. Expert Rev Vaccines 3(2):189–197

Fisher-Hoch SP, McCormick JB, Auperin D, Brown BG, Castor M, Perez G, Ruo S, Conaty A, Brammer L, Bauer S (1989) Protection of rhesus monkeys from fatal Lassa fever by vaccination with a recombinant vaccinia virus containing the Lassa virus glycoprotein gene. Proc Natl Acad Sci USA 86(1):317–321

Frame JD, Verbrugge GP, Gill RG, Pinneo L (1984) The use of Lassa fever convalescent plasma in Nigeria. Trans R Soc Trop Med Hyg 78(3):319–324

Frantz PN, Teeravechyan S, Tangy F (2018) Measles-derived vaccines to prevent emerging viral diseases. Microbes Infect 20(9–10):493–500

Furuta Y, Komeno T, Nakamura T (2017) Favipiravir (T-705), a broad spectrum inhibitor of viral RNA polymerase. Proc Jpn Acad Ser B Phys Biol Sci 93(7):449–463

Gabriel M, Adomeh DI, Ehimuan J, Oyakhilome J, Omomoh EO, Ighodalo Y, Olokor T, Bonney K, Pahlmann M, Emmerich P, Lelke M, Brunotte L, Ölschläger S, Thomé-Bolduan C, Becker-Ziaja B, Busch C, Odia I, Ogbaini-Emovon E, Okokhere PO, Okogbenin SA, Akpede GO, Schmitz H, Asogun DA, Günther S (2018) Development and evaluation of antibody-capture immunoassays for detection of Lassa virus nucleoprotein-specific immunoglobulin M and G. PLoS Negl Trop Dis 12(3):e0006361

Geisbert TW, Jones S, Fritz EA, Shurtleff AC, Geisbert JB, Liebscher R, Grolla A, Ströher U, Fernando L, Daddario KM, Guttieri MC, Mothé BR, Larsen T, Hensley LE, Jahrling PB, Feldmann H (2005) Development of a new vaccine for the prevention of Lassa fever. PLoS Med 2(6):e183

Goicochea MA, Zapata JC, Bryant J, Davis H, Salvato MS, Lukashevich IS (2012) Evaluation of Lassa virus vaccine immunogenicity in a CBA/J-ML29 mouse model. Vaccine 30(8):1445–1452

Gowen BB, Hickerson BT, York J, Westover JB, Sefing EJ, Bailey KW, Wandersee L, Nunberg JH (2021a) Second-Generation Live-Attenuated Candid#1 vaccine virus resists

reversion and protects against lethal Junín virus infection in guinea pigs. J Virol 95(14): e0039721
Gowen BB, Julander JG, London NR, Wong MH, Larson D, Morrey JD, Li DY, Bray M (2010) Assessing changes in vascular permeability in a hamster model of viral hemorrhagic fever. Virol J 7:240
Gowen BB, Naik S, Westover JB, Brown ER, Gantla VR, Fetsko A, Dagley AL, Blotter DJ, Anderson N, McCormack K, Henkel G (2021b) Potent inhibition of arenavirus infection by a novel fusion inhibitor. Antiviral Res 193:105125
Gowen BB, Wong MH, Jung KH, Sanders AB, Mendenhall M, Bailey KW, Furuta Y, Sidwell RW (2007) In vitro and in vivo activities of T-705 against arenavirus and Bunyavirus infections. Antimicrob Agents Chemother 51(9):3168–3176
Hastie KM, Zandonatti MA, Kleinfelter LM, Heinrich ML, Rowland MM, Chandran K, Branco LM, Robinson JE, Garry RF, Saphire EO (2017) Structural basis for antibody-mediated neutralization of Lassa virus. Science 356(6341):923–928
Helguera G, Jemielity S, Abraham J, Cordo SM, Martinez MG, Rodríguez JA, Bregni C, Wang JJ, Farzan M, Penichet ML, Candurra NA, Choe H (2012) An antibody recognizing the apical domain of human transferrin receptor 1 efficiently inhibits the entry of all new world hemorrhagic fever arenaviruses. J Virol 86(7):4024–4028
Heppner DG Jr, Kemp TL, Martin BK, Ramsey WJ, Nichols R, Dasen EJ, Link CJ, Das R, Xu ZJ, Sheldon EA, Nowak TA, Monath TP (2017) Safety and immunogenicity of the rVSVΔG-ZEBOV-GP Ebola virus vaccine candidate in healthy adults: a phase 1b randomised, multicentre, double-blind, placebo-controlled, dose-response study. Lancet Infect Dis 17 (8):854–866
Hulseberg CE, Fénéant L, Szymańska-de Wijs KM, Kessler NP, Nelson EA, Shoemaker CJ, Schmaljohn CS, Polyak SJ, White JM (2019) Arbidol and other low-molecular-weight drugs that inhibit Lassa and Ebola viruses. J Virol 93(8)
Jahrling PB (1983) Protection of Lassa virus-infected guinea pigs with Lassa-immune plasma of guinea pig, primate, and human origin. J Med Virol 12(2):93–102
Jahrling PB, Frame JD, Rhoderick JB, Monson MH (1985) Endemic Lassa fever in Liberia. IV. Selection of optimally effective plasma for treatment by passive immunization. Trans R Soc Trop Med Hyg 79(3):380–384
Jahrling PB, Peters CJ, Stephen EL (1984) Enhanced treatment of Lassa fever by immune plasma combined with ribavirin in cynomolgus monkeys. J Infect Dis 149(3):420–427
Jiang J, Banglore P, Cashman KA, Schmaljohn CS, Schultheis K, Pugh H, Nguyen J, Humeau LM, Broderick KE, Ramos SJ (2019) Immunogenicity of a protective intradermal DNA vaccine against Lassa virus in cynomolgus macaques. Hum Vaccin Immunother 15 (9):2066–2074
Jiang X, Dalebout TJ, Bredenbeek PJ, Carrion R Jr, Brasky K, Patterson J, Goicochea M, Bryant J, Salvato MS, Lukashevich IS (2011) Yellow fever 17D-vectored vaccines expressing Lassa virus GP1 and GP2 glycoproteins provide protection against fatal disease in guinea pigs. Vaccine 29(6):1248–1257
Johnson DM, Sokoloski KJ, Jokinen JD, Pfeffer TL, Chu YK, Adcock RS, Chung D, Tretyakova I, Pushko P, Lukashevich IS (2020) Advanced safety and genetic stability in mice of a novel DNA-launched venezuelan equine Encephalitis virus vaccine with rearranged structural genes. Vaccines 8(1)
Johnson KM, McCormick JB, Webb PA, Smith ES, Elliott LH, King IJ (1987) Clinical virology of Lassa fever in hospitalized patients. J Infect Dis 155(3):456–464
Kang CY, Luo L, Wainberg MA, Li Y (1999) Development of HIV/AIDS vaccine using chimeric Gag-Env virus-like particles. Biol Chem 380(3):353–364
Keane E, Gilles HM (1977) Lassa fever in Panguma Hospital, Sierra Leone, 1973–6. Br Med J 1 (6073):1399–1402
Kiley MP, Lange JV, Johnson KM (1979) Protection of rhesus monkeys from Lassa virus by immunisation with closely related Arenavirus. Lancet 2(8145):738

Kotturi MF, Botten J, Sidney J, Bui HH, Giancola L, Maybeno M, Babin J, Oseroff C, Pasquetto V, Greenbaum JA, Peters B, Ting J, Do D, Vang L, Alexander J, Grey H, Buchmeier MJ, Sette A (2009) A multivalent and cross-protective vaccine strategy against arenaviruses associated with human disease. PLoS Pathog 5(12):e1000695

Kurup D, Fisher CR, Scher G, Yankowski C, Testa A, Keshwara R, Abreu-Mota T, Lambert R, Ferguson M, Rinaldi W, Ruiz L, Wirblich C, Schnell MJ (2021) Tetravalent rabies-vectored filovirus and Lassa fever vaccine induces long-term immunity in nonhuman primates. J Infect Dis 224(6):995–1004

Larson RA, Dai D, Hosack VT, Tan Y, Bolken TC, Hruby DE, Amberg SM (2008) Identification of a broad-spectrum arenavirus entry inhibitor. J Virol 82(21):10768–10775

Lau JY, Tam RC, Liang TJ, Hong Z (2002) Mechanism of action of ribavirin in the combination treatment of chronic HCV infection. Hepatology 35(5):1002–1009

Leifer E, Gocke DJ, Bourne H (1970) Lassa fever, a new virus disease of man from West Africa. II. Report of a laboratory-acquired infection treated with plasma from a person recently recovered from the disease. Am J Trop Med Hyg 19(4):677–679

Li S, Sun Z, Pryce R, Parsy ML, Fehling SK, Schlie K, Siebert CA, Garten W, Bowden TA, Strecker T, Huiskonen JT (2016) Acidic pH-induced conformations and LAMP1 binding of the Lassa virus glycoprotein spike. PLoS Pathog 12(2):e1005418

Liljeström P, Garoff H (1991) A new generation of animal cell expression vectors based on the semliki forest virus replicon. Biotechnology 9(12):1356–1361

Lingas G, Rosenke K, Safronetz D, Guedj J (2021) Lassa viral dynamics in non-human primates treated with favipiravir or ribavirin. PLoS Comput Biol 17(1):e1008535

López-Camacho C, Abbink P, Larocca RA, Dejnirattisai W, Boyd M, Badamchi-Zadeh A, Wallace ZR, Doig J, Velazquez RS, Neto RDL, Coelho DF, Kim YC, Donald CL, Owsianka A, De Lorenzo G, Kohl A, Gilbert SC, Dorrell L, Mongkolsapaya J, Patel AH, Screaton GR, Barouch DH, Hill AVS, Reyes-Sandoval A (2018) Rational Zika vaccine design via the modulation of antigen membrane anchors in chimpanzee adenoviral vectors. Nat Commun 9(1):2441

Lukashevich IS (1992) Generation of reassortants between African arenaviruses. Virology 188 (2):600–605

Lukashevich IS (2012) Advanced vaccine candidates for Lassa fever. Viruses 4(11):2514–2557

Lukashevich IS, Carrion R Jr, Salvato MS, Mansfield K, Brasky K, Zapata J, Cairo C, Goicochea M, Hoosien GE, Ticer A, Bryant J, Davis H, Hammamieh R, Mayda M, Jett M, Patterson J (2008) Safety, immunogenicity, and efficacy of the ML29 reassortant vaccine for Lassa fever in small non-human primates. Vaccine 26(41):5246–5254

Lukashevich IS, Djavani M, Shapiro K, Sanchez A, Ravkov E, Nichol ST, Salvato MS (1997) The Lassa fever virus L gene: nucleotide sequence, comparison, and precipitation of a predicted 250 kDa protein with monospecific antiserum. J Gen Virol 78(Pt 3):547–551

Lukashevich IS, Patterson J, Carrion R, Moshkoff D, Ticer A, Zapata J, Brasky K, Geiger R, Hubbard GB, Bryant J, Salvato MS (2005) A live attenuated vaccine for Lassa fever made by reassortment of Lassa and Mopeia viruses. J Virol 79(22):13934–13942

Madu IG, Files M, Gharaibeh DN, Moore AL, Jung KH, Gowen BB, Dai D, Jones KF, Tyavanagimatt SR, Burgeson JR, Korth MJ, Bedard KM, Iadonato SP, Amberg SM (2018) A potent Lassa virus antiviral targets an arenavirus virulence determinant. PLoS Pathog 14(12): e1007439

Mahmood K, Bright RA, Mytle N, Carter DM, Crevar CJ, Achenbach JE, Heaton PM, Tumpey TM, Ross TM (2008) H5N1 VLP vaccine induced protection in ferrets against lethal challenge with highly pathogenic H5N1 influenza viruses. Vaccine 26(42):5393–5399

Maiztegui JI, Fernandez NJ, de Damilano AJ (1979) Efficacy of immune plasma in treatment of Argentine haemorrhagic fever and association between treatment and a late neurological syndrome. Lancet 2(8154):1216–1217

Matassov D, Mire CE, Latham T, Geisbert JB, Xu R, Ota-Setlik A, Agans KN, Kobs DJ, Wendling MQS, Burnaugh A, Rudge TL, Jr, Sabourin CL, Egan MA, Clarke DK,

Geisbert TW, Eldridge JH (2018) Single-dose trivalent VesiculoVax vaccine protects macaques from lethal Ebolavirus and marburgvirus challenge. J Virol 92(3)
Mateo M, Reynard S, Carnec X, Journeaux A, Baillet N, Schaeffer J, Picard C, Legras-Lachuer C, Allan R, Perthame E, Hillion KH, Pietrosemoli N, Dillies MA, Barrot L, Vallve A, Barron S, Fellmann L, Gaillard JC, Armengaud J, Carbonnelle C, Raoul H, Tangy F, Baize S (2019) Vaccines inducing immunity to Lassa virus glycoprotein and nucleoprotein protect macaques after a single shot. Sci Transl Med 11(512)
Mateo M, Reynard S, Journeaux A, Germain C, Hortion J, Carnec X, Picard C, Baillet N, Borges-Cardoso V, Merabet O, Vallve A, Barron S, Jourjon O, Lacroix O, Duthey A, Dirheimer M, Jouvion G, Moreau PH, Fellmann L, Carbonnelle C, Raoul H, Tangy F, Baize S (2021) A single-shot Lassa vaccine induces long-term immunity and protects cynomolgus monkeys against heterologous strains. Sci Transl Med 13(597)
McCormick JB, Fisher-Hoch SP (2002) Lassa fever. Curr Top Microbiol Immunol 262:75–109
McCormick JB, King IJ, Webb PA, Scribner CL, Craven RB, Johnson KM, Elliott LH, Belmont-Williams R (1986) Lassa fever. Effective therapy with ribavirin. N Engl J Med 314 (1):20–26
McCormick JB, Mitchell SW, Kiley MP, Ruo S, Fisher-Hoch SP (1992) Inactivated Lassa virus elicits a non protective immune response in rhesus monkeys. J Med Virol 37(1):1–7
McCormick JB, Webb PA, Krebs JW, Johnson KM, Smith ES (1987) A prospective study of the epidemiology and ecology of Lassa fever. J Infect Dis 155(3):437–444
McGettigan JP, Pomerantz RJ, Siler CA, McKenna PM, Foley HD, Dietzschold B, Schnell MJ (2003) Second-generation rabies virus-based vaccine vectors expressing human immunodeficiency virus type 1 gag have greatly reduced pathogenicity but are highly immunogenic. J Virol 77(1):237–244
McKee KT Jr, Oro JG, Kuehne AI, Spisso JA, Mahlandt BG (1993) Safety and immunogenicity of a live-attenuated Junin (Argentine hemorrhagic fever) vaccine in rhesus macaques. Am J Trop Med Hyg 48(3):403–411
Medeiros LR, Rosa DD, da Rosa MI, Bozzetti MC, Zanini RR (2009) Efficacy of human papillomavirus vaccines: a systematic quantitative review. Int J Gynecol Cancer 19(7):1166–1176
Mendenhall M, Russell A, Smee DF, Hall JO, Skirpstunas R, Furuta Y, Gowen BB (2011) Effective oral favipiravir (T-705) therapy initiated after the onset of clinical disease in a model of arenavirus hemorrhagic Fever. PLoS Negl Trop Dis 5(10):e1342
Meulen J, Badusche M, Satoguina J, Strecker T, Lenz O, Loeliger C, Sakho M, Koulemou K, Koivogui L, Hoerauf A (2004) Old and New World arenaviruses share a highly conserved epitope in the fusion domain of the glycoprotein 2, which is recognized by Lassa virus-specific human CD4+ T-cell clones. Virology 321(1):134–143
Mihailova M, Boos M, Petrovskis I, Ose V, Skrastina D, Fiedler M, Sominskaya I, Ross S, Pumpens P, Roggendorf M, Viazov S (2006) Recombinant virus-like particles as a carrier of B- and T-cell epitopes of hepatitis C virus (HCV). Vaccine 24(20):4369–4377
Mire CE, Cross RW, Geisbert JB, Borisevich V, Agans KN, Deer DJ, Heinrich ML, Rowland MM, Goba A, Momoh M, Boisen ML, Grant DS, Fullah M, Khan SH, Fenton KA, Robinson JE, Branco LM, Garry RF, Geisbert TW (2017) Human-monoclonal-antibody therapy protects nonhuman primates against advanced Lassa fever. Nat Med 23(10):1146–1149
Mire CE, Matassov D, Geisbert JB, Latham TE, Agans KN, Xu R, Ota-Setlik A, Egan MA, Fenton KA, Clarke DK, Eldridge JH, Geisbert TW (2015) Single-dose attenuated Vesiculovax vaccines protect primates against Ebola Makona virus. Nature 520(7549):688–691
Morrison HG, Bauer SP, Lange JV, Esposito JJ, McCormick JB, Auperin DD (1989) Protection of guinea pigs from Lassa fever by vaccinia virus recombinants expressing the nucleoprotein or the envelope glycoproteins of Lassa virus. Virology 171(1):179–188
Moshkoff DA, Salvato MS, Lukashevich IS (2007) Molecular characterization of a reassortant virus derived from Lassa and Mopeia viruses. Virus Genes 34(2):169–176

Müller H, Fehling SK, Dorna J, Urbanowicz RA, Oestereich L, Krebs Y, Kolesnikova L, Schauflinger M, Krähling V, Magassouba N, Fichet-Calvet E, Ball JK, Kaufmann A, Bauer S, Becker S, von Messling V, Strecker T (2020) Adjuvant formulated virus-like particles expressing native-like forms of the Lassa virus envelope surface glycoprotein are immunogenic and induce antibodies with broadly neutralizing activity. NPJ Vaccines 5(1):71
Munster VJ, Wells D, Lambe T, Wright D, Fischer RJ, Bushmaker T, Saturday G, van Doremalen N, Gilbert SC, de Wit E, Warimwe GM (2017) Protective efficacy of a novel simian adenovirus vaccine against lethal MERS-CoV challenge in a transgenic human DPP4 mouse model. NPJ Vaccines 2:28
Murawski MR, McGinnes LW, Finberg RW, Kurt-Jones EA, Massare MJ, Smith G, Heaton PM, Fraire AE, Morrison TG (2010) Newcastle disease virus-like particles containing respiratory syncytial virus G protein induced protection in BALB/c mice, with no evidence of immunopathology. J Virol 84(2):1110–1123
Neuman BW, Bederka LH, Stein DA, Ting JP, Moulton HM, Buchmeier MJ (2011) Development of peptide-conjugated morpholino oligomers as pan-arenavirus inhibitors. Antimicrob Agents Chemother 55(10):4631–4638
Ng PP, Helguera G, Daniels TR, Lomas SZ, Rodriguez JA, Schiller G, Bonavida B, Morrison SL, Penichet ML (2006) Molecular events contributing to cell death in malignant human hematopoietic cells elicited by an IgG3-avidin fusion protein targeting the transferrin receptor. Blood 108(8):2745–2754
Nikisins S, Rieger T, Patel P, Müller R, Günther S, Niedrig M (2015) International external quality assessment study for molecular detection of Lassa virus. PLoS Negl Trop Dis 9(5):e0003793
O'Hearn AE, Voorhees MA, Fetterer DP, Wauquier N, Coomber MR, Bangura J, Fair JN, Gonzalez JP, Schoepp RJ (2016) Serosurveillance of viral pathogens circulating in West Africa. Virol J 13(1):163
Oestereich L, Rieger T, Lüdtke A, Ruibal P, Wurr S, Pallasch E, Bockholt S, Krasemann S, Muñoz-Fontela C, Günther S (2016) Efficacy of Favipiravir alone and in combination with ribavirin in a lethal, immunocompetent mouse model of Lassa fever. J Infect Dis 213(6):934–938
Olschläger S, Lelke M, Emmerich P, Panning M, Drosten C, Hass M, Asogun D, Ehichioya D, Omilabu S, Günther S (2010) Improved detection of Lassa virus by reverse transcription-PCR targeting the 5′ region of S RNA. J Clin Microbiol 48(6):2009–2013
Pannetier D, Faure C, Georges-Courbot MC, Deubel V, Baize S (2004) Human macrophages, but not dendritic cells, are activated and produce alpha/beta interferons in response to Mopeia virus infection. J Virol 78(19):10516–10524
Pannetier D, Reynard S, Russier M, Journeaux A, Tordo N, Deubel V, Baize S (2011) Human dendritic cells infected with the nonpathogenic Mopeia virus induce stronger T-cell responses than those infected with Lassa virus. J Virol 85(16):8293–8306
Pécheur EI, Borisevich V, Halfmann P, Morrey JD, Smee DF, Prichard M, Mire CE, Kawaoka Y, Geisbert TW, Polyak SJ (2016) The synthetic antiviral drug Arbidol inhibits globally prevalent pathogenic viruses. J Virol 90(6):3086–3092
Perez M, Craven RC, de la Torre JC (2003) The small RING finger protein Z drives arenavirus budding: implications for antiviral strategies. Proc Natl Acad Sci USA 100(22):12978–12983
Plewe MB, Gantla VR, Sokolova NV, Shin YJ, Naik S, Brown ER, Fetsko A, Zhang L, Kalveram B, Freiberg AN, Henkel G, McCormack K (2021) Discovery of a novel highly potent broad-spectrum heterocyclic chemical series of arenavirus cell entry inhibitors. Bioorg Med Chem Lett 41:127983
Purushotham J, Lambe T, Gilbert SC (2019) Vaccine platforms for the prevention of Lassa fever. Immunol Lett 215:1–11
Pushko P, Geisbert J, Parker M, Jahrling P, Smith J (2001) Individual and bivalent vaccines based on alphavirus replicons protect guinea pigs against infection with Lassa and Ebola viruses. J Virol 75(23):11677–11685

Pushko P, Kort T, Nathan M, Pearce MB, Smith G, Tumpey TM (2010) Recombinant H1N1 virus-like particle vaccine elicits protective immunity in ferrets against the 2009 pandemic H1N1 influenza virus. Vaccine 28(30):4771–4776

Pushko P, Parker M, Ludwig GV, Davis NL, Johnston RE, Smith JF (1997) Replicon-helper systems from attenuated Venezuelan equine encephalitis virus: expression of heterologous genes in vitro and immunization against heterologous pathogens in vivo. Virology 239(2):389–401

Pushko P, Tumpey TM, Van Hoeven N, Belser JA, Robinson R, Nathan M, Smith G, Wright DC, Bright RA (2007) Evaluation of influenza virus-like particles and Novasome adjuvant as candidate vaccine for avian influenza. Vaccine 25(21):4283–4290

Raabe V, Koehler J (2017) Laboratory diagnosis of Lassa fever. J Clin Microbiol 55(6):1629–1637

Raabe VN, Kann G, Ribner BS, Morales A, Varkey JB, Mehta AK, Lyon GM, Vanairsdale S, Faber K, Becker S, Eickmann M, Strecker T, Brown S, Patel K, De Leuw P, Schuettfort G, Stephan C, Rabenau H, Klena JD, Rollin PE, McElroy A, Ströher U, Nichol S, Kraft CS, Wolf T (2017) Favipiravir and ribavirin treatment of epidemiologically linked cases of Lassa fever. Clin Infect Dis 65(5):855–859

Regules JA, Beigel JH, Paolino KM, Voell J, Castellano AR, Hu Z, Muñoz P, Moon JE, Ruck RC, Bennett JW, Twomey PS, Gutiérrez RL, Remich SA, Hack HR, Wisniewski ML, Josleyn MD, Kwilas SA, Van Deusen N, Mbaya OT, Zhou Y, Stanley DA, Jing W, Smith KS, Shi M, Ledgerwood JE, Graham BS, Sullivan NJ, Jagodzinski LL, Peel SA, Alimonti JB, Hooper JW, Silvera PM, Martin BK, Monath TP, Ramsey WJ, Link CJ, Lane HC, Michael NL, Davey RT Jr, Thomas SJ (2017) A recombinant vesicular stomatitis virus Ebola vaccine. N Engl J Med 376(4):330–341

Robinson JE, Hastie KM, Cross RW, Yenni RE, Elliott DH, Rouelle JA, Kannadka CB, Smira AA, Garry CE, Bradley BT, Yu H, Shaffer JG, Boisen ML, Hartnett JN, Zandonatti MA, Rowland MM, Heinrich ML, Martínez-Sobrido L, Cheng B, de la Torre JC, Andersen KG, Goba A, Momoh M, Fullah M, Gbakie M, Kanneh L, Koroma VJ, Fonnie R, Jalloh SC, Kargbo B, Vandi MA, Gbetuwa M, Ikponmwosa O, Asogun DA, Okokhere PO, Follarin OA, Schieffelin JS, Pitts KR, Geisbert JB, Kulakoski PC, Wilson RB, Happi CT, Sabeti PC, Gevao SM, Khan SH, Grant DS, Geisbert TW, Saphire EO, Branco LM, Garry RF (2016) Most neutralizing human monoclonal antibodies target novel epitopes requiring both Lassa virus glycoprotein subunits. Nat Commun 7:11544

Rosenke K, Feldmann H, Westover JB, Hanley PW, Martellaro C, Feldmann F, Saturday G, Lovaglio J, Scott DP, Furuta Y, Komeno T, Gowen BB, Safronetz D (2018) Use of Favipiravir to treat Lassa virus infection in macaques. Emerg Infect Dis 24(9):1696–1699

Safronetz D, Mire C, Rosenke K, Feldmann F, Haddock E, Geisbert T, Feldmann H (2015a) A recombinant vesicular stomatitis virus-based Lassa fever vaccine protects guinea pigs and macaques against challenge with geographically and genetically distinct Lassa viruses. PLoS Negl Trop Dis 9(4):e0003736

Safronetz D, Rosenke K, Westover JB, Martellaro C, Okumura A, Furuta Y, Geisbert J, Saturday G, Komeno T, Geisbert TW, Feldmann H, Gowen BB (2015b) The broad-spectrum antiviral Favipiravir protects guinea pigs from lethal Lassa virus infection post-disease onset. Sci Rep 5:14775

Sahin M, Remy MM, Merkler D, Pinschewer DD (2021) The Janus Kinase inhibitor ruxolitinib prevents terminal shock in a mouse model of arenavirus hemorrhagic fever. Microorganisms 9 (3)

Salvato MS, Domi A, Guzmán-Cardozo C, Medina-Moreno S, Zapata JC, Hsu H, McCurley N, Basu R, Hauser M, Hellerstein M, Guirakhoo F (2019) A single dose of modified vaccinia ankara expressing lassa virus-like particles protects mice from lethal intra-cerebral virus challenge. Pathogens 8(3)

Sanchez A, Pifat DY, Kenyon RH, Peters CJ, McCormick JB, Kiley MP (1989) Junin virus monoclonal antibodies: characterization and cross-reactivity with other arenaviruses. J Gen Virol 70(Pt 5):1125–1132

Satterly NG, Voorhees MA, Ames AD, Schoepp RJ (2017) Comparison of MagPix assays and enzyme-linked immunosorbent assay for detection of hemorrhagic fever viruses. J Clin Microbiol 55(1):68–78

Shiraki K, Daikoku T (2020) Favipiravir, an anti-influenza drug against life-threatening RNA virus infections. Pharmacol Ther 209:107512

Spence JS, Melnik LI, Badani H, Wimley WC, Garry RF (2014) Inhibition of arenavirus infection by a glycoprotein-derived peptide with a novel mechanism. J Virol 88(15):8556–8564

Spohn G, Jennings GT, Martina BE, Keller I, Beck M, Pumpens P, Osterhaus AD, Bachmann MF (2010) A VLP-based vaccine targeting domain III of the West Nile virus E protein protects from lethal infection in mice. Virol J 7:146

Stadhouders R, Pas SD, Anber J, Voermans J, Mes TH, Schutten M (2010) The effect of primer-template mismatches on the detection and quantification of nucleic acids using the 5' nuclease assay. J Mol Diagn 12(1):109–117

Stein DR, Warner BM, Soule G, Tierney K, Frost KL, Booth S, Safronetz D (2019) A recombinant vesicular stomatitis-based Lassa fever vaccine elicits rapid and long-term protection from lethal Lassa virus infection in guinea pigs. NPJ Vaccines 4:8

Stylianou E, Harrington-Kandt R, Beglov J, Bull N, Pinpathomrat N, Swarbrick GM, Lewinsohn DA, Lewinsohn DM, McShane H (2018) Identification and evaluation of novel protective antigens for the development of a candidate tuberculosis subunit vaccine. Infect Immun 86(7)

Swenson DL, Warfield KL, Negley DL, Schmaljohn A, Aman MJ, Bavari S (2005) Virus-like particles exhibit potential as a pan-filovirus vaccine for both Ebola and Marburg viral infections. Vaccine 23(23):3033–3042

Thompson JM, Whitmore AC, Staats HF, Johnston R (2008) The contribution of type I interferon signaling to immunity induced by alphavirus replicon vaccines. Vaccine 26(39):4998–5003

Torriani G, Trofimenko E, Mayor J, Fedeli C, Moreno H, Michel S, Heulot M, Chevalier N, Zimmer G, Shrestha N, Plattet P, Engler O, Rothenberger S, Widmann C, Kunz S (2019) Identification of clotrimazole derivatives as specific inhibitors of arenavirus fusion. J Virol 93 (6)

Trombley AR, Wachter L, Garrison J, Buckley-Beason VA, Jahrling J, Hensley LE, Schoepp RJ, Norwood DA, Goba A, Fair JN, Kulesh DA (2010) Comprehensive panel of real-time TaqMan polymerase chain reaction assays for detection and absolute quantification of filoviruses, arenaviruses, and New World hantaviruses. Am J Trop Med Hyg 82(5):954–960

Uckun FM, Pendergrass S, Venkatachalam TK, Qazi S, Richman D (2002) Stampidine is a potent inhibitor of Zidovudine- and nucleoside analog reverse transcriptase inhibitor-resistant primary clinical human immunodeficiency virus type 1 isolates with thymidine analog mutations. Antimicrob Agents Chemother 46(11):3613–3616

Uckun FM, Petkevich AS, Vassilev AO, Tibbles HE, Titov L (2004) Stampidine prevents mortality in an experimental mouse model of viral hemorrhagic fever caused by lassa virus. BMC Infect Dis 4:1

Vander Veen RL, Harris DL, Kamrud KI (2012) Alphavirus replicon vaccines. Anim Health Res Rev 13(1):1–9

Warfield KL, Swenson DL, Olinger GG, Kalina WV, Aman MJ, Bavari S (2007) Ebola virus-like particle-based vaccine protects nonhuman primates against lethal Ebola virus challenge. J Infect Dis 196(Suppl 2):S430-437

Warimwe GM, Gesharisha J, Carr BV, Otieno S, Otingah K, Wright D, Charleston B, Okoth E, Elena LG, Lorenzo G, Ayman EB, Alharbi NK, Al-dubaib MA, Brun A, Gilbert SC, Nene V, Hill AV (2016) Chimpanzee adenovirus vaccine provides multispecies protection against rift valley fever. Sci Rep 6:20617

Warner BM, Safronetz D, Stein DR (2018) Current research for a vaccine against Lassa hemorrhagic fever virus. Drug Des Devel Ther 12:2519–2527

Warren TK, Wells J, Panchal RG, Stuthman KS, Garza NL, Van Tongeren SA, Dong L, Retterer CJ, Eaton BP, Pegoraro G, Honnold S, Bantia S, Kotian P, Chen X, Taubenheim BR, Welch LS, Minning DM, Babu YS, Sheridan WP, Bavari S (2014) Protection against filovirus diseases by a novel broad-spectrum nucleoside analogue BCX4430. Nature 508(7496):402–405

Witkowski JT, Robins RK, Sidwell RW, Simon LN (1972) Design, synthesis, and broad spectrum antiviral activity of 1-D-ribofuranosyl-1,2,4-triazole-3-carboxamide and related nucleosides. J Med Chem 15(11):1150–1154

Woolsey C, Cross RW, Agans KN, Borisevich V, Deer DJ, Geisbert JB, Gerardi C, Latham TE, Fenton KA, Egan MA, Eldridge JH, Geisbert TW, Matassov D (2022) A highly attenuated Vesiculovax vaccine rapidly protects nonhuman primates against lethal Marburg virus challenge. PLoS Negl Trop Dis 16(5):e0010433

Xiong C, Levis R, Shen P, Schlesinger S, Rice CM, Huang HV (1989) Sindbis virus: an efficient, broad host range vector for gene expression in animal cells. Science 243(4895):1188–1191

Zapata JC, Poonia B, Bryant J, Davis H, Ateh E, George L, Crasta O, Zhang Y, Slezak T, Jaing C, Pauza CD, Goicochea M, Moshkoff D, Lukashevich IS, Salvato MS (2013) An attenuated Lassa vaccine in SIV-infected rhesus macaques does not persist or cause arenavirus disease but does elicit Lassa virus-specific immunity. Virol J 10:52

Zeitlin L, Cross RW, Geisbert JB, Borisevich V, Agans KN, Prasad AN, Enterlein S, Aman MJ, Bornholdt ZA, Brennan MB, Campbell L, Kim D, Mlakar N, Moyer CL, Pauly MH, Shestowsky W, Whaley KJ, Fenton KA, Geisbert TW (2021) Therapy for argentine hemorrhagic fever in nonhuman primates with a humanized monoclonal antibody. Proc Natl Acad Sci USA 118(11)

Zeitlin L, Geisbert JB, Deer DJ, Fenton KA, Bohorov O, Bohorova N, Goodman C, Kim D, Hiatt A, Pauly MH, Velasco J, Whaley KJ, Altmann F, Gruber C, Steinkellner H, Honko AN, Kuehne AI, Aman MJ, Sahandi S, Enterlein S, Zhan X, Enria D, Geisbert TW (2016) Monoclonal antibody therapy for Junin virus infection. Proc Natl Acad Sci USA 113 (16):4458–4463

Lassa Virus Structural Biology and Replication

Robert F. Garry

Contents

Abstract Lassa virus (LASV) is the causative agent of Lassa fever, an often-fatal hemorrhagic fever that is endemic in West Africa. LASV virions are enveloped and contain two single-stranded RNA genome segments. Both segments are ambisense and encode two proteins. The nucleoprotein associates with viral RNAs forming ribonucleoprotein complexes. The glycoprotein complex mediates viral attachment and entry. The Zinc protein serves as the matrix protein. Large is a polymerase that catalyzes viral RNA transcription and replication. LASV virion entry occurs via a

R. F. Garry (✉)
School of Medicine, Department of Microbiology and Immunology, Tulane University, New Orleans, LA 70112, USA
e-mail: rfgarry@tulane.edu

Zalgen Labs, Frederick, MD 21703, USA

Global Virus Network (GVN), Baltimore, MD 21201, USA

Current Topics in Microbiology and Immunology (2023) 440: 147–164
https://doi.org/10.1007/82_2023_262

Published Online: 27 April 2023

clathrin-independent endocytic pathway usually involving alpha-dystroglycan and lysosomal associated membrane protein 1 as surface and intracellular receptors, respectively. Advances in understanding LASV structural biology and replication have facilitated development of promising vaccine and drug candidates.

1 Introduction

Lassa virus (LASV) is the prototype of the *Arenaviridae* family (subfamily *Mammarenavirus*, Order *Bunyavirales*). It is the causative agent of Lassa fever, an often-fatal hemorrhagic fever. Lassa fever is a classic example of a zoonosis (Bell-Kareem and Smither 2021; Smither and Bell-Kareem 2021). Most human infections are due to exposure to LASV-containing excretions from *Mastomys natalensis* (multimammate mouse), other rodent reservoirs or intermediate hosts (Fichet-Calvet and Rogers 2009; Lecompte et al. 2006; Mylne et al. 2015; Ogbu et al. 2007; Olayemi et al. 2016). Hundreds of thousands of LASV Infections with several thousand Lassa fever deaths are estimated to occur each year in West Africa, mostly in Nigeria, Sierra Leone, Liberia and Guinea. The Lassa fever endemic zone appears to be expanding, which may reflect in part expansion of the range of its rodent reservoirs (Happi et al. 2022; Klitting et al. 2022; Sogoba et al. 2012). Case fatality rates (CFR) for Lassa fever vary from approximately 25% in Nigeria to 70% in Sierra Leone (Okokhere et al. 2018; Shaffer et al. 2014). A 2015–2016 surge in Nigeria was associated with a 60% CFR (Buba et al. 2018). Further large geographically dispersed outbreaks have occurred in Nigeria in subsequent years (Siddle et al. 2018). The disease is more severe in pregnant women with a CFR of up to 90% in the third trimester (Branco et al. 2011; Price et al. 1988).

While LASV has extreme genetic divergence (Andersen et al. 2015; Klitting et al. 2020), virion architectures, protein structures and functions, and replication strategies appear similar across its multiple lineages (Fig. 1). LASV virions have a lipid-containing envelope and contain two single-stranded RNA segments (Fig. 1A). Structures derived from host cell ribosomes are also present. Both the small (S) and large (L) RNA segments are ambisense; one portion is positive sense, the same sense as messenger RNA (mRNA), and another portion is negative sense, the complement of mRNA (Auperin et al. 1986) (Fig. 1B). Each segment encodes two proteins (Oldstone 2002). Nucleoprotein (NP) and Large (L) mRNAs are transcribed from the negative-sense portions of genome-sense S sand L segments, respectively. The Glycoprotein Complex (GPC) and Zinc protein (Z) are expressed from mRNAs transcribed from S and L antigenomic RNAs, respectively. Here, the structural features of LASV proteins and their roles in viral replication are discussed.

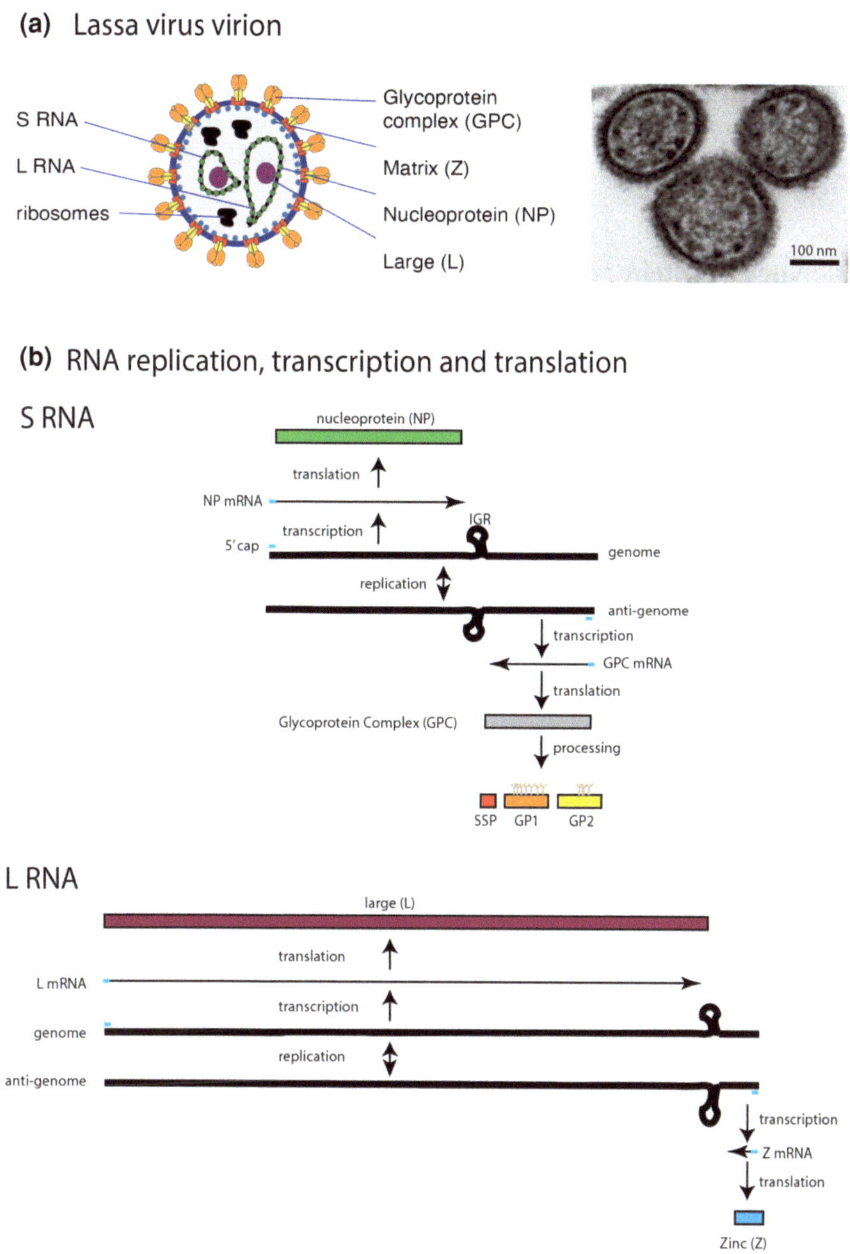

Fig. 1 Lassa virus virion architecture, RNA replication and protein expression. Panel A: The LASV virion contains host ribosomes and two segments of single-stranded RNA. Electron micrograph of LASV from Mali is reproduced from (Safronetz et al. 2010). Panel B: Both LASV genome segments are ambisense. Each segment encodes two viral proteins. A stem–loop structure between the positive and negative sense genes terminates the bidirectional transcription. Replication of genome segments involves an antigenome intermediate

2 LASV Proteins

NP, Z and GPC play important structural roles in the virion, as well as other roles in virus replication or pathogenesis. L is also present in virions at a limited number of copies and is involved in viral genome replication and mRNA transcription.

3 Nucleoprotein

NP, the most abundant LASV protein in virions and infected cells, associates with genomic and antigenomic RNAs forming ribonucleoprotein (RNP) complexes (Figs. 1A and 2). NP association with L is essential both for transcription of viral messenger RNAs (mRNAs) and replication of genome segments for incorporation into progeny virions (Hastie et al. 2011a, 2011b, 2012b). NP also interacts with Z and contributes to the packaging of RNPs during virion morphogenesis. It has multiple roles in suppressing antiviral responses (Garry 2022).

NP is organized into distinct amino-terminal and carboxy-terminal domains connected by a flexible linker (Fig. 2A). The N-terminal domain contains a crevice that binds RNA (Fig. 2B) (Hastie et al. 2011b). The crevice is lined with basic amino acids that interact with the negatively charged backbone of viral RNAs (Fig. 2B). Steps in the oligomeric assembly of NP and the bound genomic RNA within the ribonucleoprotein complex (RNP) are not well defined. The structure of the RNP In virions is presumed to resemble the beads-on-a-string morphology assumed by RNP complexes of other RNA viruses (Young and Howard 1983). Full-length LASV NP (Qi et al. 2011) crystallizes as a trimer, but in a conformation that is unable to bind RNA (Hastie et al. 2011b). Therefore, conformational changes or rearrangements of NP must occur during RNP assembly.

The C-terminal domain of NP has striking structural similarity to DEDD superfamily exonucleases with major structural elements showing only minor deviations when aligned (Hastie et al. 2011a, 2012b). NP shares the signature Asp-Glu-Asp-Asp (DEDD) catalytic residues in the active site with other DEDD superfamily members (Fig. 2C). The presence of a histidine (H) or tyrosine, (Y) near the DEDD active site subdivides the superfamily into DEDDh or DEDDy groups with NP possessing a His (26). Degradation of small viral doubled-stranded RNAs by NP blunts the type I interferon (IFN-I)-mediated immune response and blocks activation of the innate immune system (Hastie et al. 2012a). NP also prevents nuclear translocation of nuclear factor kappa B (NF-κB) via association with the retinoic acid-inducible gene I (RIG-I) and I-kappa-B kinase epsilon (IKKε), thereby inhibiting the activation and nuclear translocation of interferon regulatory factor 3 (IRF-3) (Garry 2022; Zinzula and Tramontano 2013).

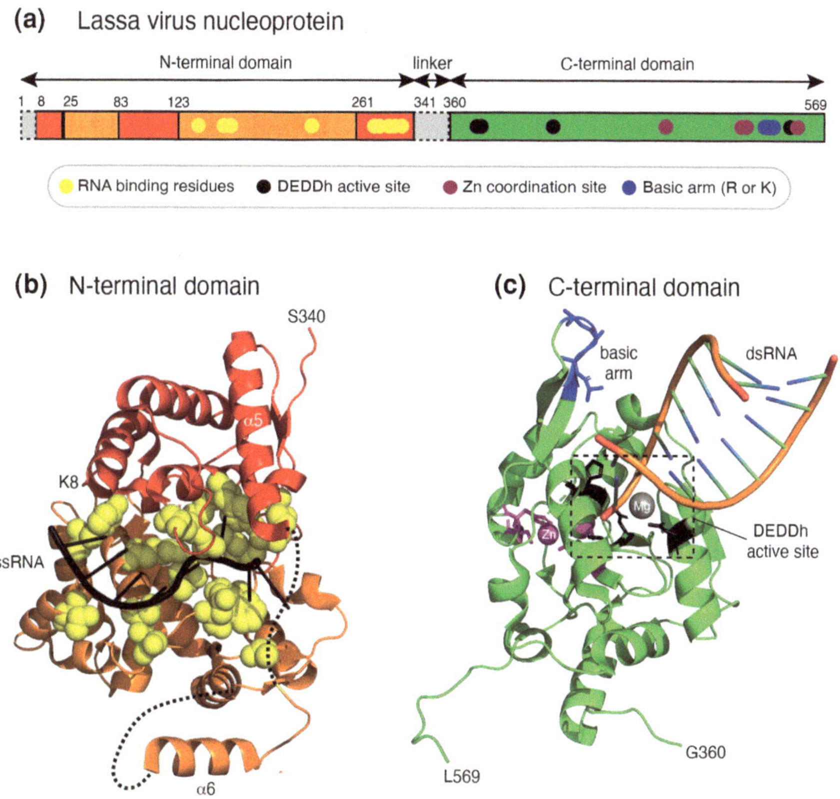

Fig. 2 Structure of the Lassa virus nucleoprotein. NP has N- and C-terminal domains connected by a flexible linker. The DEDDh active site includes D389, E391, D466, D533, and H528. The basic arm includes residues K516, K517, K518, and R519. K = Lysine R = Arginine. A magnesium (Mn) ion in the active site is coordinated to D389, E391, and D533. Panel B: Structure of the RNA-binding domain of Lassa virus NP in complex with ssRNA (Hastie et al. 2011b). Panel C: structure of the C-terminal, immunosuppressive domain of LASV NP in complex with double-stranded RNA (Hastie et al. 2011a)

4 Glycoprotein Complex

GPC is the only viral protein present on the LASV virion surface (Figs. 1A and 3). GPC binds cell surface and intracellular receptors and catalyzes fusion of the virion envelope with intracellular vesicle membranes permitting entry of the genome segments into the cytoplasm (Oppliger et al. 2016; Pasqual et al. 2011). Translation of GPC mRNA produces a glycoprotein precursor that is cleaved by signal peptidase (SPase) and subtilisin kexin isozyme-1/site-1 (SKI-1) protease (Fig. 3A) (Hastie et al. 2017). The cleavage products remain associated in the virion, although GP1 is shed from infected cells (Branco et al. 2010b). Thus, virion GPC is comprised of three components: glycoprotein 1 (GP1), the receptor binding subunit,

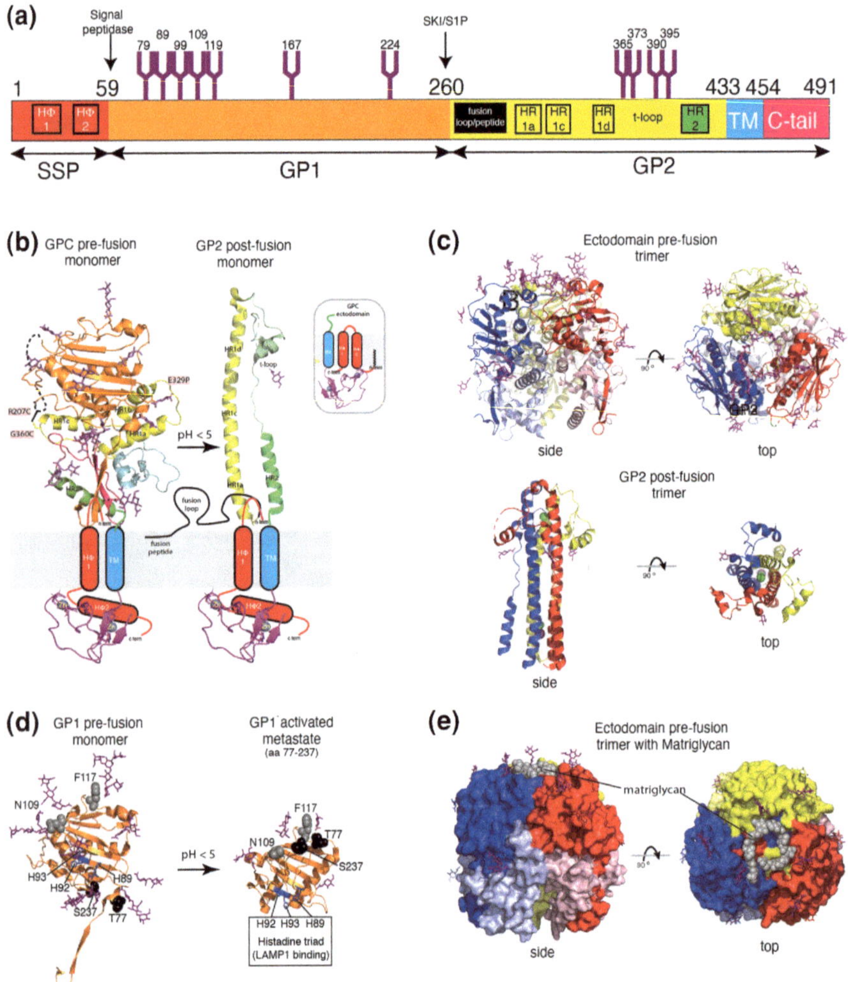

Fig. 3 Structure of the Lassa virus glycoprotein complex. Panel A: GPC is comprised of three components: glycoprotein 1 (GP1), glycoprotein 2 (GP2), and a virion-associated stable signal peptide (SSP). A glycoprotein precursor is cleaved by signal peptidase (SPase) and subtilisin kexin isozyme-1/site-1 (SKI-1) protease. Panel B. GP1, GP2 and SSP remain associated in GPC monomers on the virion surface. LASV glycoprotein structure from (Hastie et al. 2017). Exposure of GPC to low pH results in structural rearrangement of GP2 (Zhang et al. 2019). Structure of the Junin virus intravirion domain (Briknarova et al. 2011), which shares numerous conserved amino acids with the LASV GP2 intravirion domain is shown. Inset shows an alternative orientation of SSP: Amino terminal: N-term. Carboxyl terminal: c-term. Panel C: GPC is arranged as a trimer on the virion surface. Panel D. Exposure of GPC to low pH results in structural rearrangement of GP1. GP1 structure at low pH is from (Cohen-Dvashi et al. 2015). Panel E: the trimeric structure binds matriglycan a component of alpha-dystroglycan (α-DG). Trimeric structure binding matriglycan is from (Li et al. 2022)

glycoprotein 2 (GP2), a class I membrane fusion protein, and a virion-associated stable signal peptide (SSP) (Fig. 3B). GPC is arranged as a trimer on the virion surface (Fig. 3C) (Hastie et al. 2017). More than 30 glycans are added to the GP1 and 2 subunits of the trimer (Hastie et al. 2017). Both GP1 (Fig. 3D) and GP2 (Fig. 3B) undergo structural arrangements upon exposure to the low pH environments encountered during the multistep entry of the virion via the endocytic route. The trimeric structure is necessary to bind matriglycan a component of alpha-dystroglycan (α-DG), the primary LASV cell surface receptor (Fig. 3E) (Li et al. 2022).

GP2 includes an intravirion (intracellular) domain characterized by the presence of several His and cysteine (Cys) residues. Six of these residues (H455, H457, C463, H467, C475, C477; numbered according to the LASV Josiah strain) are conserved throughout mammarenaviruses (York and Nunberg 2006, 2007). The unusually long SSP is essential for LASV infectivity and appears to contribute to trimeric stability via interactions with the transmembrane (TM) and intracellular domains of GP2. By analogy to the JUNV GP2 (York and Nunberg 2007), it is likely that the LASV GP2 intracellular domain coordinates two Zinc ions (Fig. 3B). One or more Cys of LASV SSP may to be involved in forming the second Zinc binding site in coordination with HIs and Cys residues of the GP2 intracellular domain.

Data regarding location, orientation, and structural interactions of SSP with GP2 are incomplete and conflicting. SSPs of LASV and other mammarenaviruses have two hydrophobic domains (Fig. 3A). Previous work by Nunberg and coworkers (Briknarova et al. 2011; York and Nunberg 2006; York et al. 2004) supports a model of SSP in which the hydrophobic domains form a hairpin in the virion membrane with both the N-terminus and C-terminus in the interior (Fig. 3B inset). Additionally, the N-terminus of arenavirus SSP has been shown to be myristylated (York and Nunberg 2016; York et al. 2004). In contrast, Katz and coworkers (Katz et al. 2022) provided cryo-electron microscopy (cryo-EM) evidence showing that the second hydrophobic region of SSP is located inside the virion. These data suggest that SSP undergoes a topological rearrangement after SPase cleavage that reorients the first hydrophobic region of SSP domain within the membrane such that its C-terminus is exterior.

5 Zinc Binding Protein

Z is necessary and sufficient for LASV budding, functioning as the viral matrix proteins (Figs. 1 and 4) (Branco et al. 2010a; Perez et al. 2003; Urata and Yasuda 2015). This multifunctional protein also negatively regulates viral replication and transcription via interaction with L as well as genome incorporation into virions through interactions with NP (Shao et al. 2018). Z is important for the suppression of both viral and host cell translation via interaction with the cellular translation initiation factor eIF4E (Shao et al. 2018).

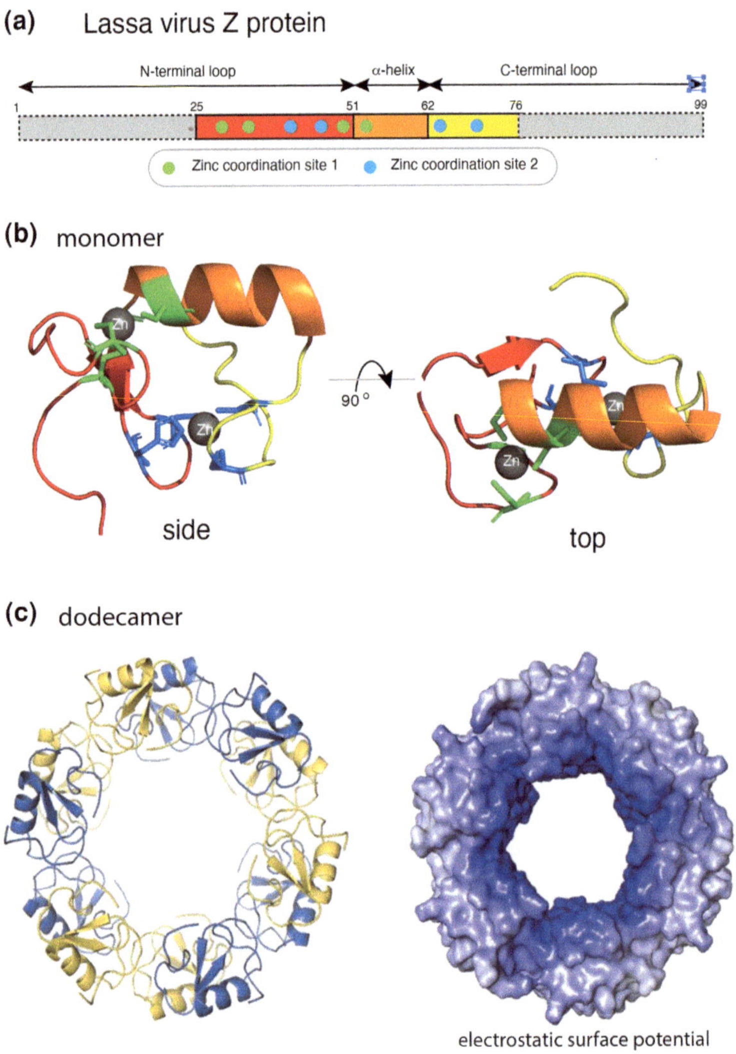

Fig. 4 Structure of the Lassa virus Zinc-binding protein. Panel A: the structure of LASV Z monomer Z consists of an amino terminal (N-terminal) loop, an alpha helix (α-helix) and a carboxyl-terminal (C-terminal) loop that has an antiparallel beta sheets. Zinc site 1: C31, C34, C50, C53. Zinc site 2: C44, H47, C64, C67. Cysteine: C. Histidine: H. Gray boxes are residues not resolved in the crystal structures. Panel B: structure of Z monomer as determined by Hastie et al. 2016 (Hastie et al. 2016). Panel C: cartoon depiction and surface structures of the Z dodecamer. Reproduced from (Hastie et al. 2016). In the surface presentation blue is positive electrostatic surface potential. Negative would have been red but is not present

LASV Z occurs in monomeric and oligomeric forms (Fig. 4) (Hastie et al. 2016; Volpon et al. 2008). Each Z monomer contains two zinc-binding sites, an alpha (α) helix, two antiparallel beta (β) strands, and two loop regions (Fig. 4A). The α-helix and the residues surrounding the two Zinc atoms were resolved in the Z crystal structure (Fig. 4B). However, residues corresponding to the N- and C-terminal portions of LASV Z are not seen (Fig. 4A) (Hastie et al. 2016). These portions of Z are highly flexible and fast exchanging by deuterium exchange mass spectroscopy, which explains why they could not be resolved by X-ray crystallography (Hastie et al. 2016). Z dimers are formed by hydrophobic interactions between the α-helices of two monomers (Fig. 4C). A dodecamer of Z monomer forms in a concentration and time dependent manner; six Z dimers associate to form a donut-shaped ring. The interior of this 12-membered ring is highly basic. It is unclear if the Z dodecameric ring is a building block for LASV virions. A Z mutant L56R, which is unable to form the dodecamer, is released into the supernatant. Analytical ultracentrifugation studies of LCMV Z demonstrate that it forms 24-member "bodies" (Kentsis et al. 2002). It is possible that LASV Z forms other distinct structures each of which may mediate a distinct function.

6 Large

L plays a central role in viral replication (Figs. 1 and 5), including replication of viral genome segments and transcription of viral mRNAs (Hass et al. 2008, 2006; Lelke et al. 2010a; Reguera et al. 2016; Vogel et al. 2019). Because of its large size L is not amenable to X-ray crystallographic analyses. However, advances in cryo-EM enabled 3-dimentional structures of L to be determined to high resolution (Kouba et al. 2021). These structures revealed that LASV L shares structural similarities to polymerases of certain other RNA viruses. For example, the influenza virus polymerase is a heterotrimeric protein complex comprising proteins designated PA, PB1 and PB2. LASV L can be divided into PA-like, PB1-like and PB2-like regions (Fig. 5A). The PA-like region has an N-terminal endonuclease (Lelke et al. 2010b). The endonuclease domain is followed by an extended linker that wraps around the polymerase core that connects to a prominent feature, denoted the pyramid domain. At the beginning of the pyramid domain, there is a zinc-binding site. However, this site is specific to the LASV strain used for the structural analyses and is not conserved in other LASV lineages. The PB1-like region possesses the RNA-dependent RNA polymerase (RdRp) activity. Like certain other RNA virus polymerases, the structure of the catalytic regions of L resembles a cupped right hand. The PB-1-like region contains the canonical fingers and palm and a catalytic triad of aspartates. The PB2-like region completes the hand with a thumb domain and a "thumb-ring" as well as a putative cap-binding domain.

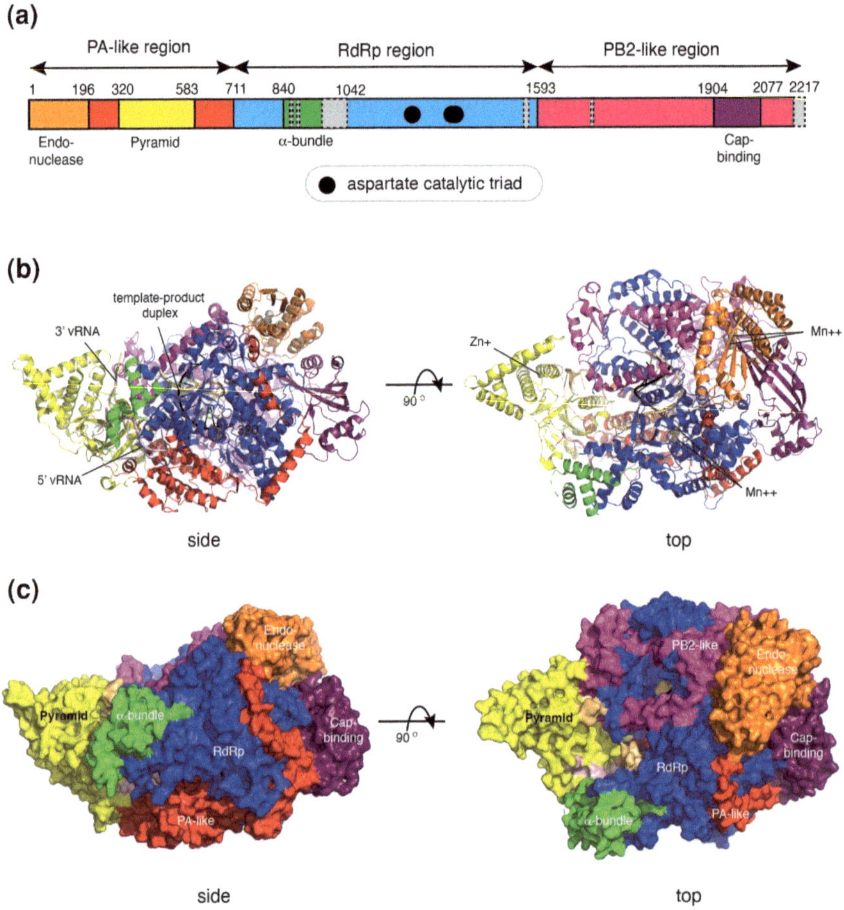

Fig. 5 Structure of the Lassa virus large protein. Panel A: LASV L consists of PA-1 like, PB1-like and PB2-like regions. Gray boxes are residues not resolved in the crystal structures. D1190, D1331 and D1332 are key residues in the active site. Aspartic acid = D Panel B: cartoon depiction of the Elongation structure of LASV L as determined by (Kouba et al. 2021). Panel C: surface depiction of L structure

7 LASV Replication

LASV's replication strategy follows a well-established pattern for enveloped viruses: attachment, entry, uncoating, RNA replication and protein expression, assembly, and budding/egress.

8 Attachment

Lassa virus uses extracellular and intracellular receptors for entry into host cells (Fig. 6) (Cohen-Dvashi et al. 2015; Cohen-Dvashi et al. 2016; Enriquez et al. 2022; Hulseberg et al. 2018; Kunz et al. 2002; Oldstone and Campbell 2011; Rojek and Kunz 2008). Dystroglycan (DG) is composed of two subunits: α-DG is extracellular, highly glycosylated, and the major LASV surface receptor and α-DG that serves as the transmembrane anchor (Fig. 6, step 1). LASV binding to α-DG depends on matriglycan, which is linked to α-DG through phosphorylated *O*-mannose (Li et al. 2022; Rojek et al. 2007). Matriglycan comprises repeating units of xylose (Xyl) and glucuronic acid (GlcA) that extend large distances from the cell surface. Like-acetylglucosaminyltransferase (LARGE) is required for the attachment of ligand-binding moieties to α-DG via phosphorylated *O*-mannose (Yoshida-Moriguchi et al. 2010). Binding of LASV GP to β-DG induces tyrosine phosphorylation of β-DG's cytosolic domain and triggers β-DG's dissociation from the cytoskeletal adaptor utrophin. This dissociation may facilitate endocytosis of the virus–receptor complex.

Certain cell-types deficient in α-DG, such as hepatocytes and mice lacking a functional LARGE are susceptible to LASV infection. This suggested that LASV can utilize other entry factors besides α-DG. Studies by Kawaoka and colleagues identified the Tyro3/Axl/Mer (TAM) receptor tyrosine kinases Axl and Tyro3/Dtk as potential LASV receptor candidates (Shimojima et al. 2012). TAM family members play a role in entry of several RNA viruses. T-cell immunoglobulin mucin I (TIM-1) (Brouillette et al. 2018), dendritic cell-specific intercellular adhesion molecule-3 grabbing nonintegrin (DC-SIGN) and liver and lymph node sinusoidal endothelial calcium-dependent lectin (LSECtin) have also been identified as potential LASV cell surface receptors (Shimojima et al. 2012).

9 Entry

After α-DG binding, LASV enters the host cell via a clathrin-independent endocytic process (Fig. 6, step 2) (Rojek et al. 2008). Following entry in an uncoated vesicle that has an internal pH of 7–8, fusion with multivesicular bodies lowers the intravesicular pH to approximately 6 (Oppliger et al. 2016). The acidic pH induces a change in GP1 configuration that releases the virion from α-DG (Li et al. 2016). Transport of the virion to late endosomal compartments that have a pH of approximately 5 induces a further conformational changein GP1 to a primed metastate that allows binding to lysosome-associated membrane protein 1 (LAMP1) (Fig. 6, step 3) (Hastie and Saphire 2018; Hastie et al. 2017; Hulseberg et al. 2018; Israeli et al. 2017; Zhang et al. 2022). A histidine triad (H92, H93, and H230) located on the β-sheet face of GP1 is important both for pH sensing and LAMP1 binding (Fig. 3D). In the neutral-pH, prefusion complex, H92 packs against N89.

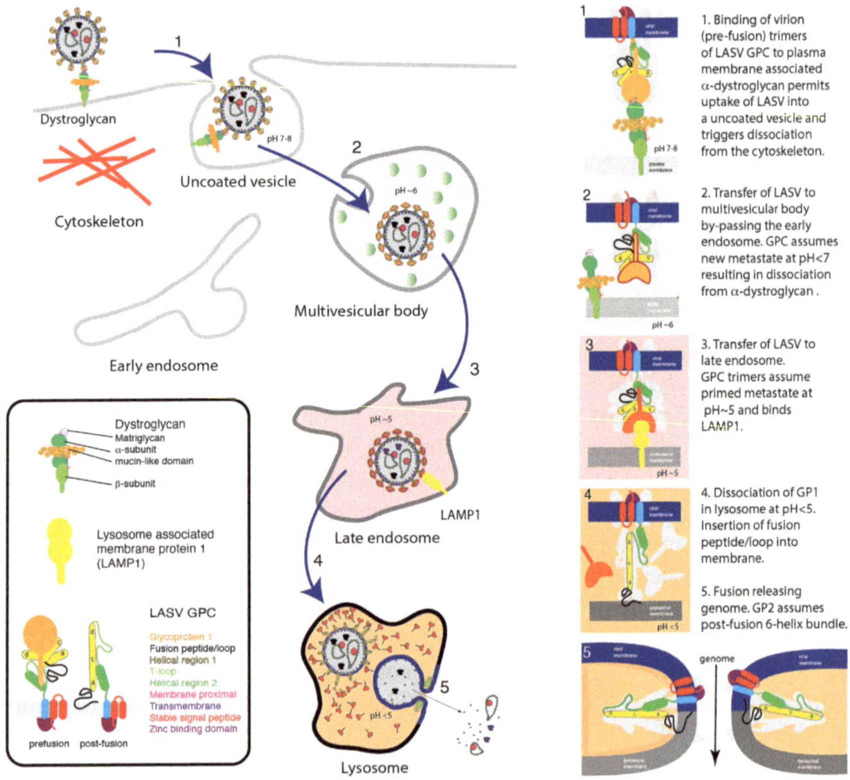

Fig. 6 Lassa virus attachment and entry. LASV virions bind to cell surface and intracellular receptors to enable entry of the viral genome into the cell cytoplasm

In the low-pH structure of GP1, H92 is in essentially the same location, but N89, is shifted ~10 Å away. Additionally, at neutral-pH H93 is located on the underside of a β sheet, but at low-pH it is on top. Amino acids threonine 77 and Serine 237 of the low pH form of GP1 are oriented upward, away from GP2. Thus, helices, loops, and the termini of GP1 appear to be flexible and adopt alternate positions when not associated with GP2. Low pH alter stability of GP1 subunit permitting its release from GP2. Entry into late endosome that have a pH below 5 induces another dramatic rearrangement, the GP2 subunit arranges into a 6-helix bundle (Figs. 3B and C, 6, step 4 and 5). Fusion between the virion membrane and the vesicular membrane permits entry of the viral genome segments into the cell cytoplasm (Fig. 6, step 5) (Hastie and Saphire 2018).

10 RNA Replication and Translation

Transcription from and replication of the small and large genome segments occurs upon their entry into the cytoplasm. Each segment is a template for the synthesis of two different types of RNA products: capped viral mRNAs for the NP and L from S and L, respectively, as well as noncapped full-length complementary RNA (cRNA) copies or antigenomes (Fig. 1B). The antigenomes are templates for the synthesis of genome copies that are incorporated into progeny virions, as well as transcription of mRNAs for GPC and Z. Transcription of all LASV mRNAs is initiated using a capped primer derived from host mRNA by a yet to be elucidated "cap-snatching" mechanism. Cap-snatching likely involves the endonuclease activity and putative cap-binding domain of L (Reguera et al. 2016). In this scenario endonuclease cleaves 5′ ends of cellular mRNAs, which then primes synthesis of mRNAs resulting in viral mRNAs with several host-derived nucleotides following the 5' cap. A caveat is that studies with the isolated C-terminal region of LASV L have yet to detect cap-binding activity, readily found in other viral polymerases (Lehmann et al. 2014). Additional studies are needed to determine how—of if—a functional cap-binding site is present on LASV L.

11 Virion Assembly and Egress

Z controls the packing and budding of infectious particles (Djavani et al. 1997; Perez et al. 2003; Yasuda and Jiro 2012). Z protein alone can produce LASV virus-like particles (VLPs) (Branco et al. 2010a; Urata et al. 2006). Z controls the assembly of NP, M, L and the genomic RNA and the release of LASV virions from the cell membrane. LASV possesses electron dense features within the virion when visualized by electron microscopy. It is presumed that these features originate from cellular ribosomes (Murphy and Whitfield 1975; Rowe et al. 1970). Proteomics analyses confirmed that LASV VLPs incorporate ribosomal proteins; proteins of both large and small ribosomal subunits were detected (Gale et al. 2019). Cytoskeletal, chaperone, and nucleic acid-binding proteins were also detected in LASV VLPs. Efforts to identify catalytic activity of Lassa VLP-associated ribosomes were unsuccessful suggesting that the ribosome-like structures in LASV virions are either incomplete or immature.

12 Conclusion

LASV is a high priority pathogen causing severe hemorrhagic fever across a broad endemic range in West Africa. It is also a potential bioterrorism agent. Novel approaches to treat and prevent infection with LASV are urgently need for public health and biodefense (Melnik 2022). Treatment is currently limited to off-label use of the antiviral drug ribavirin. There are no FDA-approved vaccines against LASV. Advances in the understanding of Lassa virus structural biology and replication have greatly facilitated the development of candidate LASV vaccines (Bernasconi et al. 2020; Gouglas et al. 2019). Several candidates are in the early stages of clinical evaluation. Several candidate antiviral drugs, such a human monoclonal antibodies (Li et al. 2022; Mire et al. 2017; Robinson et al. 2016), are also in advanced stages of preclinical or clinical development.

References

Andersen KG, Shapiro BJ, Matranga CB, Sealfon R, Lin AE, Moses LM, Folarin OA, Goba A, Odia I, Ehiane PE, Momoh M, England EM, Winnicki S, Branco LM, Gire SK, Phelan E, Tariyal R, Tewhey R, Omoniwa O, Fullah M, Fonnie R, Fonnie M, Kanneh L, Jalloh S, Gbakie M, Saffa S, Karbo K, Gladden AD, Qu J, Stremlau M, Nekoui M, Finucane HK, Tabrizi S, Vitti JJ, Birren B, Fitzgerald M, McCowan C, Ireland A, Berlin AM, Bochicchio J, Tazon-Vega B, Lennon NJ, Ryan EM, Bjornson Z, Milner DA Jr, Lukens AK, Broodie N, Rowland M, Heinrich M, Akdag M, Schieffelin JS, Levy D, Akpan H, Bausch DG, Rubins K, McCormick JB, Lander ES, Gunther S, Hensley L, Okogbenin S, Schaffner SF, Okokhere PO, Khan SH, Grant DS, Akpede GO, Asogun DA, Gnirke A, Levin JZ, Happi CT, Garry RF, Sabeti PC (2015) Clinical sequencing uncovers origins and evolution of lassa virus. Cell 162:738–750

Auperin DD, Sasso DR, McCormick JB (1986) Nucleotide sequence of the glycoprotein gene and intergenic region of the Lassa virus S genome RNA. Virology 154:155–167

Bell-Kareem AR, Smither AR (2021) Epidemiology of lassa fever. Curr Top Microbiol Immunol

Bernasconi V, Kristiansen PA, Whelan M, Roman RG, Bettis A, Yimer SA, Gurry C, Andersen SR, Yeskey D, Mandi H, Kumar A, Holst J, Clark C, Cramer JP, Rottingen JA, Hatchett R, Saville M, Norheim G (2020) Developing vaccines against epidemic-prone emerging infectious diseases. Bundesgesundheitsblatt Gesundheitsforschung Gesundheitsschutz 63:65–73

Branco LM, Boisen ML, Andersen KG, Grove JN, Moses LM, Muncy IJ, Henderson LA, Schieffellin JS, Robinson JE, Bangura JJ, Grant DS, Raabe VN, Fonnie M, Zaitsev EM, Sabeti PC, Garry RF (2011) Lassa hemorrhagic fever in a late term pregnancy from northern Sierra Leone with a positive maternal outcome: case report. Virol J 8:404

Branco LM, Grove JN, Geske FJ, Boisen ML, Muncy IJ, Magliato SA, Henderson LA, Schoepp RJ, Cashman KA, Hensley LE, Garry RF (2010a) Lassa virus-like particles displaying all major immunological determinants as a vaccine candidate for Lassa hemorrhagic fever. Virol J 7:279

Branco LM, Grove JN, Moses LM, Goba A, Fullah M, Momoh M, Schoepp RJ, Bausch DG, Garry RF (2010b) Shedding of soluble glycoprotein 1 detected during acute Lassa virus infection in human subjects. Virol J 7:306

Briknarova K, Thomas CJ, York J, Nunberg JH (2011) Structure of a zinc-binding domain in the Junin virus envelope glycoprotein. J Biol Chem 286:1528–1536

Brouillette RB, Phillips EK, Patel R, Mahauad-Fernandez W, Moller-Tank S, Rogers KJ, Dillard JA, Cooney AL, Martinez-Sobrido L, Okeoma C, Maury W (2018) TIM-1 mediates dystroglycan-independent entry of lassa virus. J Virol 92

Buba MI, Dalhat MM, Nguku PM, Waziri N, Mohammad JO, Bomoi IM, Onyiah AP, Onwujei J, Balogun MS, Bashorun AT, Nsubuga P, Nasidi A (2018) Mortality among confirmed lassa fever cases during the 2015–2016 outbreak in Nigeria. Am J Public Health 108:262–264

Cohen-Dvashi H, Cohen N, Israeli H, Diskin R (2015) Molecular mechanism for LAMP1 recognition by Lassa virus. J Virol 89:7584–7592
Cohen-Dvashi H, Israeli H, Shani O, Katz A, Diskin R (2016) The role of LAMP1 binding and pH sensing by the spike complex of Lassa virus. J Virol
Djavani M, Lukashevich IS, Sanchez A, Nichol ST, Salvato MS (1997) Completion of the Lassa fever virus sequence and identification of a RING finger open reading frame at the L RNA 5' end. Virology 235:414–418
Enriquez AS, Buck TK, Li H, Norris MJ, Moon-Walker A, Zandonatti MA, Harkins SS, Robinson JE, Branco LM, Garry RF, Saphire EO, Hastie KM (2022) Delineating the mechanism of anti-Lassa virus GPC-A neutralizing antibodies. Cell Rep 39:110841
Fichet-Calvet E, Rogers DJ (2009) Risk maps of Lassa fever in West Africa. PLoS Negl Trop Dis 3:e388
Gale TV, Horton TM, Hoffmann AR, Branco LM, Garry RF (2019) Host proteins identified in extracellular viral particles as targets for broad-spectrum antiviral inhibitors. J Proteome Res 18:7–17
Garry RF (2022) Lassa fever—the road ahead. Nature reviews. Microbiology, 1–10
Gouglas D, Christodoulou M, Plotkin SA, Hatchett R (2019) CEPI: driving progress towards epidemic preparedness and response. Epidemiol Rev
Happi AN, Olumade TJ, Ogunsanya OA, Sijuwola AE, Ogunleye SC, Oguzie JU, Nwofoke C, Ugwu CA, Okoro SJ, Otuh PI, Ngele LN, Ojo OO, Adelabu A, Adeleye RF, Oyejide NE, Njaka CS, Heeney JL, Happi CT (2022) Increased prevalence of Lassa fever virus-positive rodents and diversity of infected species found during human Lassa fever epidemics in Nigeria. Microbiol Spectr 10:e0036622
Hass M, Lelke M, Busch C, Becker-Ziaja B, Günther S (2008) Mutational evidence for a structural model of the Lassa virus RNA polymerase domain and identification of two residues, Gly1394 and Asp1395, that are critical for transcription but not replication of the genome. J Virol 82:10207–10217
Hass M, Westerkofsky M, Müller S, Becker-Ziaja B, Busch C, Günther S (2006) Mutational analysis of the Lassa virus promoter. J Virol 80:12414–12419
Hastie KM, Bale S, Kimberlin CR, Saphire EO (2012a) Hiding the evidence: two strategies for innate immune evasion by hemorrhagic fever viruses. Curr Opin Virol 2:151–156
Hastie KM, Kimberlin CR, Zandonatti MA, MacRae IJ, Saphire EO (2011a) Structure of the Lassa virus nucleoprotein reveals a dsRNA-specific 3' to 5' exonuclease activity essential for immune suppression. Proc Natl Acad Sci USA 108:2396–2401
Hastie KM, King LB, Zandonatti MA, Saphire EO (2012b) Structural basis for the dsRNA specificity of the Lassa virus NP exonuclease. PLoS ONE 7:e44211
Hastie KM, Liu T, Li S, King LB, Ngo N, Zandonatti MA, Woods VL Jr, de la Torre JC, Saphire EO (2011b) Crystal structure of the Lassa virus nucleoprotein-RNA complex reveals a gating mechanism for RNA binding. Proc Natl Acad Sci USA 108:19365–19370
Hastie KM, Saphire EO (2018) Lassa virus glycoprotein: stopping a moving target. Current Opinion Virol
Hastie KM, Zandonatti M, Liu T, Li S, Woods VL Jr, Saphire EO (2016) Crystal structure of the oligomeric form of Lassa virus matrix protein Z. J Virol 90:4556–4562
Hastie KM, Zandonatti MA, Kleinfelter LM, Heinrich ML, Rowland MM, Chandran K, Branco LM, Robinson JE, Garry RF, Saphire EO (2017) Structural basis for antibody-mediated neutralization of Lassa virus. Science 356:923–928
Hulseberg CE, Feneant L, Szymanska KM, White JM (2018) Lamp1 increases the efficiency of lassa virus infection by promoting fusion in less acidic endosomal compartments. mBio 9
Israeli H, Cohen-Dvashi H, Shulman A, Shimon A, Diskin R (2017) Mapping of the Lassa virus LAMP1 binding site reveals unique determinants not shared by other old world arenaviruses. PLoS Pathog 13:e1006337
Katz M, Weinstein J, Eilon-Ashkenazy M, Gehring K, Cohen-Dvashi H, Elad N, Fleishman SJ, Diskin R (2022) Structure and receptor recognition by the Lassa virus spike complex. Nature 603:174–179

Kentsis A, Gordon RE, Borden KL (2002) Self-assembly properties of a model RING domain. Proc Natl Acad Sci USA 99:667–672
Klitting R, Kafetzopoulou LE, Thiery W, Dudas G, Gryseels S, Kotamarthi A, Vrancken B, Gangavarapu K, Momoh M, Sandi JD, Goba A, Alhasan F, Grant DS, Okogbenin S, Ogbaini-Emovo E, Garry RF, Smither AR, Zeller M, Pauthner MG, McGraw M, Hughes LD, Duraffour S, Günther S, Suchard MA, Lemey P, Andersen KG, Dellicour S (2022) Predicting the evolution of the Lassa virus endemic area and population at risk over the next decades. Nat Commun 13:5596
Klitting R, Mehta SB, Oguzie JU, Oluniyi PE, Pauthner MG, Siddle KJ, Andersen KG, Happi CT, Sabeti PC (2020) Lassa virus genetics. Curr Top Microbiol Immunol
Kouba T, Vogel D, Thorkelsson SR, Quemin ERJ, Williams HM, Milewski M, Busch C, Günther S, Grünewald K, Rosenthal M, Cusack S (2021) Conformational changes in Lassa virus L protein associated with promoter binding and RNA synthesis activity. Nat Commun 12:7018
Kunz S, Borrow P, Oldstone MB (2002) Receptor structure, binding, and cell entry of arenaviruses. Curr Top Microbiol Immunol 262:111–137
Lecompte E, Fichet-Calvet E, Daffis S, Koulemou K, Sylla O, Kourouma F, Dore A, Soropogui B, Aniskin V, Allali B, Kouassi Kan S, Lalis A, Koivogui L, Gunther S, Denys C, ter Meulen J (2006) Mastomys natalensis and Lassa fever, West Africa. Emerg Infect Dis 12:1971–1974
Lehmann M, Pahlmann M, Jérôme H, Busch C, Lelke M, Günther S (2014) Role of the C terminus of Lassa virus L protein in viral mRNA synthesis. J Virol 88:8713–8717
Lelke M, Brunotte L, Busch C, Gunther S (2010a) An N-terminal region of Lassa virus L protein plays a critical role in transcription but not replication of the virus genome. J Virol 84:1934–1944
Lelke M, Brunotte L, Busch C, Günther S (2010b) An N-terminal region of Lassa virus L protein plays a critical role in transcription but not replication of the virus genome. J Virol 84:1934–1944
Li H, Buck T, Zandonatti M, Yin J, Moon-Walker A, Fang J, Koval A, Heinrich ML, Rowland MM, Diaz Avalos R, Schendel SL, Parekh D, Zyla D, Enriquez A, Harkins S, Sullivan B, Smith V, Chukwudozie O, Watanabe R, Robinson JE, Garry RF, Branco LM, Hastie KM, Saphire EO (2022) A cocktail of protective antibodies subverts the dense glycan shield of Lassa virus. Sci Transl Med 14:eabq0991
Li S, Sun Z, Pryce R, Parsy ML, Fehling SK, Schlie K, Siebert CA, Garten W, Bowden TA, Strecker T, Huiskonen JT (2016) Acidic pH-induced conformations and LAMP1 binding of the Lassa virus glycoprotein spike. PLoS Pathog 12:e1005418
Melnik LI (2022) Lassa virus countermeasures. Curr Top Microbiol Immunol
Mire CE, Cross RW, Geisbert JB, Borisevich V, Agans KN, Deer DJ, Heinrich ML, Rowland MM, Goba A, Momoh M, Boisen ML, Grant DS, Fullah M, Khan SH, Fenton KA, Robinson JE, Branco LM, Garry RF, Geisbert TW (2017) Human-monoclonal-antibody therapy protects nonhuman primates against advanced Lassa fever. Nat Med 23:1146–1149
Murphy FA, Whitfield SG (1975) Morphology and morphogenesis of arenaviruses. Bull World Health Organ 52:409–419
Mylne AQ, Pigott DM, Longbottom J, Shearer F, Duda KA, Messina JP, Weiss DJ, Moyes CL, Golding N, Hay SI (2015) Mapping the zoonotic niche of Lassa fever in Africa. Trans R Soc Trop Med Hyg 109:483–492
Ogbu O, Ajuluchukwu E, Uneke CJ (2007) Lassa fever in West African sub-region: an overview. J Vector Borne Dis 44:1–11
Okokhere P, Colubri A, Azubike C, Iruolagbe C, Osazuwa O, Tabrizi S, Chin E, Asad S, Ediale E, Rafiu M, Adomeh D, Odia I, Atafo R, Aire C, Okogbenin S, Pahlman M, Becker-Ziaja B, Asogun D, Fradet T, Fry B, Schaffner SF, Happi C, Akpede G, Gunther S, Sabeti PC (2018) Clinical and laboratory predictors of Lassa fever outcome in a dedicated treatment facility in Nigeria: a retrospective, observational cohort study. Lancet Infect Dis 18:684–695
Olayemi A, Cadar D, Magassouba N, Obadare A, Kourouma F, Oyeyiola A, Fasogbon S, Igbokwe J, Rieger T, Bockholt S, Jerome H, Schmidt-Chanasit J, Garigliany M, Lorenzen S,

Igbahenah F, Fichet JN, Ortsega D, Omilabu S, Gunther S, Fichet-Calvet E (2016) New hosts of the Lassa virus. Sci Rep 6:25280
Oldstone MB (2002) Arenaviruses. I. The epidemiology molecular and cell biology of arenaviruses. Introduction. Curr Top Microbiol Immunol 262:V–XII
Oldstone MB, Campbell KP (2011) Decoding arenavirus pathogenesis: essential roles for alpha-dystroglycan-virus interactions and the immune response. Virology 411:170–179
Oppliger J, Torriani G, Herrador A, Kunz S (2016) Lassa virus cell entry via dystroglycan involves an unusual pathway of macropinocytosis. J Virol 90:6412–6429
Pasqual G, Rojek JM, Masin M, Chatton JY, Kunz S (2011) Old world arenaviruses enter the host cell via the multivesicular body and depend on the endosomal sorting complex required for transport. PLoS Pathog 7:e1002232
Perez M, Craven RC, de la Torre JC (2003) The small RING finger protein Z drives arenavirus budding: implications for antiviral strategies. Proc Natl Acad Sci U S A 100:12978–12983
Price ME, Fisher-Hoch SP, Craven RB, McCormick JB (1988) A prospective study of maternal and fetal outcome in acute Lassa fever infection during pregnancy. BMJ 297:584–587
Qi X, Lan S, Wang W, Schelde LM, Dong H, Wallat GD, Ly H, Liang Y, Dong C (2011) Cap binding and immune evasion revealed by Lassa nucleoprotein structure. Nature 468:779–783
Reguera J, Gerlach P, Rosenthal M, Gaudon S, Coscia F, Gunther S, Cusack S (2016) Comparative structural and functional analysis of bunyavirus and arenavirus cap-snatching endonucleases. PLoS Pathog 12:e1005636
Robinson JE, Hastie KM, Cross RW, Yenni RE, Elliott DH, Rouelle JA, Kannadka CB, Smira AA, Garry CE, Bradley BT, Yu H, Shaffer JG, Boisen ML, Hartnett JN, Zandonatti MA, Rowland MM, Heinrich ML, Martinez-Sobrido L, Cheng B, de la Torre JC, Andersen KG, Goba A, Momoh M, Fullah M, Gbakie M, Kanneh L, Koroma VJ, Fonnie R, Jalloh SC, Kargbo B, Vandi MA, Gbetuwa M, Ikponmwosa O, Asogun DA, Okokhere PO, Follarin OA, Schieffelin JS, Pitts KR, Geisbert JB, Kulakoski PC, Wilson RB, Happi CT, Sabeti PC, Gevao SM, Khan SH, Grant DS, Geisbert TW, Saphire EO, Branco LM, Garry RF (2016) Most neutralizing human monoclonal antibodies target novel epitopes requiring both Lassa virus glycoprotein subunits. Nat Commun 7:11544
Rojek JM, Campbell KP, Oldstone MB, Kunz S (2007) Old World arenavirus infection interferes with the expression of functional alpha-dystroglycan in the host cell. Mol Biol Cell 18:4493–4507
Rojek JM, Kunz S (2008) Cell entry by human pathogenic arenaviruses. Cell Microbiol 10:828–835
Rojek JM, Sanchez AB, Nguyen NT, de la Torre JC, Kunz S (2008) Different mechanisms of cell entry by human-pathogenic old world and new world arenaviruses. J Virol 82:7677–7687
Rowe WP, Murphy FA, Bergold GH, Casals J, Hotchin J, Johnson KM, Lehmann-Grube F, Mims CA, Traub E, Webb PA (1970) Arenoviruses: proposed name for a newly defined virus group. J Virol 5:651–652
Safronetz D, Lopez JE, Sogoba N, Traore SF, Raffel SJ, Fischer ER, Ebihara H, Branco L, Garry RF, Schwan TG, Feldmann H (2010) Detection of Lassa virus, Mali. Emerg Infect Dis 16:1123–1126
Shaffer JG, Grant DS, Schieffelin JS, Boisen ML, Goba A, Hartnett JN, Levy DC, Yenni RE, Moses LM, Fullah M, Momoh M, Fonnie M, Fonnie R, Kanneh L, Koroma VJ, Kargbo K, Ottomassathien D, Muncy IJ, Jones AB, Illick MM, Kulakosky PC, Haislip AM, Bishop CM, Elliot DH, Brown BL, Zhu H, Hastie KM, Andersen KG, Gire SK, Tabrizi S, Tariyal R, Stremlau M, Matschiner A, Sampey DB, Spence JS, Cross RW, Geisbert JB, Folarin OA, Happi CT, Pitts KR, Geske FJ, Geisbert TW, Saphire EO, Robinson JE, Wilson RB, Sabeti PC, Henderson LA, Khan SH, Bausch DG, Branco LM, Garry RF (2014) Lassa fever in post-conflict Sierra Leone. PLoS Negl Trop Dis 8:e2748
Shao J, Liang Y, Ly H (2018) Roles of arenavirus Z protein in mediating virion budding, viral transcription-inhibition and interferon-beta suppression. Methods Mol Biol 1604:217–227
Shimojima M, Stroher U, Ebihara H, Feldmann H, Kawaoka Y (2012) Identification of cell surface molecules involved in dystroglycan-independent Lassa virus cell entry. J Virol 86:2067–2078

Siddle KJ, Eromon P, Barnes KG, Mehta S, Oguzie JU, Odia I, Schaffner SF, Winnicki SM, Shah RR, Qu J, Wohl S, Brehio P, Iruolagbe C, Aiyepada J, Uyigue E, Akhilomen P, Okonofua G, Ye S, Kayode T, Ajogbasile F, Uwanibe J, Gaye A, Momoh M, Chak B, Kotliar D, Carter A, Gladden-Young A, Freije CA, Omoregie O, Osiemi B, Muoebonam EB, Airende M, Enigbe R, Ebo B, Nosamiefan I, Oluniyi P, Nekoui M, Ogbaini-Emovon E, Garry RF, Andersen KG, Park DJ, Yozwiak NL, Akpede G, Ihekweazu C, Tomori O, Okogbenin S, Folarin OA, Okokhere PO, MacInnis BL, Sabeti PC, Happi CT (2018) Genomic analysis of Lassa virus during an increase in cases in Nigeria in 2018. N Engl J Med 379:1745–1753

Smither AR, Bell-Kareem AR (2021) Ecology of Lassa Virus. Curr Top Microbiol Immunol

Sogoba N, Feldmann H, Safronetz D (2012) Lassa fever in West Africa: evidence for an expanded region of endemicity. Zoonoses Public Health 59(Suppl 2):43–47

Urata S, Noda T, Kawaoka Y, Yokosawa H, Yasuda J (2006) Cellular factors required for Lassa virus budding. J Virol 80:4191–4195

Urata S, Yasuda J (2015) Cis- and cell-type-dependent trans-requirements for Lassa virus-like particle production. J Gen Virol 96:1626–1635

Vogel D, Rosenthal M, Gogrefe N, Reindl S, Gunther S (2019) Biochemical characterization of the Lassa virus L protein. J Biol Chem 294:8088–8100

Volpon L, Osborne MJ, Borden KL (2008) NMR assignment of the arenaviral protein Z from Lassa fever virus. Biomol NMR Assign 2:81–84

Yasuda SU, Jiro (2012) Molecular mechanism of arenavirus assembly and budding. Viruses 4:2049–2079

York J, Nunberg JH (2006) Role of the stable signal peptide of Junin arenavirus envelope glycoprotein in pH-dependent membrane fusion. J Virol 80:7775–7780

York J, Nunberg JH (2007) A novel zinc-binding domain is essential for formation of the functional Junin virus envelope glycoprotein complex. J Virol 81:13385–13391

York J, Nunberg JH (2016) Myristoylation of the arenavirus envelope glycoprotein stable signal peptide is critical for membrane fusion but dispensable for virion morphogenesis. J Virol 90:8341–8350

York J, Romanowski V, Lu M, Nunberg JH (2004) The signal peptide of the Junin arenavirus envelope glycoprotein is myristoylated and forms an essential subunit of the mature G1–G2 complex. J Virol 78:10783–10792

Yoshida-Moriguchi T, Yu L, Stalnaker SH, Davis S, Kunz S, Madson M, Oldstone MB, Schachter H, Wells L, Campbell KP (2010) O-mannosyl phosphorylation of alpha-dystroglycan is required for laminin binding. Science 327:88–92

Young PR, Howard CR (1983) Fine structure analysis of Pichinde virus nucleocapsids. J Gen Virol 64(Pt 4):833–842

Zhang X, Wang C, Chen B, Wang Q, Xu W, Ye S, Jiang S, Zhu Y, Zhang R (2019) Crystal structure of refolding fusion core of Lassa virus GP2 and design of lassa virus fusion inhibitors. Frontiers Microbiol 10

Zhang Y, Carlos de la Torre J, Melikyan GB (2022) Human LAMP1 accelerates Lassa virus fusion and potently promotes fusion pore dilation upon forcing viral fusion with non-endosomal membrane. PLoS Pathog 18:e1010625

Zinzula L, Tramontano E (2013) Strategies of highly pathogenic RNA viruses to block dsRNA detection by RIG-I-like receptors: hide, mask, hit. Antiviral Res 100:615–635

Lassa Fever Natural History and Clinical Management

Donald S. Grant, Robert J. Samuels, Robert F. Garry, and John S. Schieffelin

Contents

D. S. Grant · R. J. Samuels
Lassa Fever Program, Kenema Government Hospital, Ministry of Health, Kenema, Sierra Leone

D. S. Grant
College of Medicine and Allied Health Sciences (COMAHS), University of Sierra Leone, Freetown, Sierra Leone

R. F. Garry
School of Medicine, Department of Microbiology and Immunology, Tulane University, New Orleans, LA 70112, USA

Zalgen Labs, Frederick, MD 21703, USA

Global Virus Network (GVN), Baltimore, MD 21201, USA

J. S. Schieffelin (✉)
School of Medicine, Department of Pediatrics, Tulane University, New Orleans, LA 70112, USA
e-mail: jschieff@tulane.edu

Current Topics in Microbiology and Immunology (2023) 440: 165–192
https://doi.org/10.1007/82_2023_263

Published Online: 28 April 2023

Abstract Lassa fever is caused by Lassa virus (LASV), an Old World Mammarenavirus that is carried by *Mastomys natalensis* and other rodents. It is endemic in Sierra Leone, Nigeria, and other countries in West Africa. The clinical presentation of LASV infection is heterogenous varying from an inapparent or mild illness to a fatal hemorrhagic fever. Exposure to LASV is usually through contact with rodent excreta. After an incubation period of 1–3 weeks, initial symptoms such as fever, headache, and fatigue develop that may progress to sore throat, retrosternal chest pain, conjunctival injection, vomiting, diarrhea, and abdominal pain. Severe illness, including hypotension, shock, and multiorgan failure, develops in a minority of patients. Patient demographics and case fatality rates are distinctly different in Sierra Leone and Nigeria. Laboratory diagnosis relies on the detection of LASV antigens or genomic RNA. LASV-specific immunoglobulin G and M assays can also contribute to clinical management. The mainstay of treatment for Lassa fever is supportive care. The nucleoside analog ribavirin is commonly used to treat acute Lassa fever but is considered useful only if treatment is begun early in the disease course. Drugs in development, including a monoclonal antibody cocktail, have the potential to impact the management of Lassa fever.

1 Introduction

Lassa fever, a hemorrhagic disease, is a significant cause of morbidity and mortality in West Africa (Garry 2023; McCormick and Fisher-Hoch 2002). The etiological agent is Lassa virus (LASV), an Old World Mammarenavirus (family: *Arenaviridae*, Order: *Bunyavirales*) (Abudurexiti et al. 2019; Oldstone 2002). Most infections result from exposure to excreta of the primary LASV reservoir, *Mastomys natalensis* (multimammate mouse) via contaminated food or household items (Bell-Kareem and Smither 2021; Lecompte et al. 2006; Smither and Bell-Kareem 2021). Household members or caregivers, including healthcare workers, can be infected by contact with the bodily fluids of an infected person (Fisher-Hoch et al. 1985, 1995; Ijarotimi et al. 2018).

Exposure to LASV in West Africa is common, particularly in rural forested regions of Nigeria and Sierra Leone (Fichet-Calvet and Rogers 2009; Monath 2019; Monath et al. 1974). In some high-risk communities, overall seroprevalence rates for LASV are greater than 30% (Grant et al. 2023b). A meta-analysis suggested an overall seroprevalence rate in West Africa of 8.7% (Kenmoe et al. 2020). A machine-learning algorithm trained on rodent reservoir ecology and human serosurveys was used to estimate that nearly 900,000 new human LASV infections occur each year in West Africa accounting for 18,000 deaths (Basinski et al. 2021). The range of LASV appears to be expanding with cases being recently identified in Benin, Cote d'Ivoire, and Mali (Klitting et al. 2022; Sogoba et al. 2012). Confirmed Lassa fever cases and deaths have increased in Nigeria since 2017, although increased ascertainment due to heightened awareness may account for some of the apparent rise (Dalhat et al. 2022; Siddle et al. 2018). Lassa fever associated with international travel has also been reported (Kofman et al. 2019; Njuguna et al. 2022a, b; Wolf et al. 2020). Such exported cases have led, albeit rarely, to secondary transmissions.

The clinical presentation of Lassa fever is extremely heterogenous, overlapping with several febrile illnesses that are common in West Africa (Ashcroft et al. 2022; Boisen et al. 2015). Here, we describe the natural history of Lassa fever and discuss current clinical practices and challenges in case management.

2 Clinical Spectrum of Lassa Virus Infection

One of the persisting mysteries regarding the natural history of Lassa fever is the extreme range of disease severity (Akpede et al. 2019; Frame 1989). Reasons for the heterogeneity in the clinical spectrum are unknown, but may relate to route and dose of infection, underlying comorbid illnesses, previous exposure to LASV (Shaffer et al. 2014, 2019, 2021), LASV strain variability (Safronetz et al. 2013; Stein et al. 2021), and human genetic predisposition (Sabeti et al. 2006). Approximately 80% of LASV infections are asymptomatic or sufficiently mild such that medical care is not sought (WHO 2022). Patients develop fever, headache, and fatigue after an incubation period of 6 to 21 days (Fig. 1). Symptoms may progress to sore throat, retrosternal chest pain, conjunctival injection, vomiting, diarrhea, and abdominal pain. Severe illness, including hypotension and shock, develops in a minority of patients often as early as the end of the first week of illness. Neurological symptoms such as altered consciousness, seizures, and tremors are present in some patients, and hearing loss is a common neurologic deficit (Ibekwe et al. 2011; Mateer et al. 2018; Okokhere et al. 2018b, 2009; Simmons 1993). A capillary leak syndrome characterized by head and neck edema, hypotension, and shock has been described. (Fisher-Hoch and McCormick 1987; Fisher-Hoch et al. 1987). Overt hemorrhage, including bleeding from the mouth, nose, vagina, and rectum, occurs in a minority of adult hospitalized cases and portends a fatal outcome (Grant et al. 2023a; Ilori et al. 2019; McCormick et al. 1987; Schieffelin et al. 2023; Shaffer et al. 2014). Among survivors, recovery is generally observed in the second and third weeks after the onset of the illness. While previous studies have shown that the onset and duration of principal signs and symptoms occur in a predictable pattern, some patients progress faster than others (McCormick et al. 1987). In a large case series, no association was found between the time from symptom onset to the presentation (Grant et al. 2023a). This observation has important implications for the design and interpretation of therapeutic trials.

As for adults, children may present with a variety of clinical manifestations following LASV infection (Adetunji et al. 2021; Akhiwu et al. 2018; Akpede et al. 2018a; Samuels et al. 2020; Webb et al. 1986). Lassa fever may present as "swollen baby syndrome" characterized by abdominal distension, bleeding, and diffuse edema (Monson et al. 1987). This syndrome mostly affects neonates, but occasionally is seen in infants and toddlers. 37% of LASV antigen-positive Sierra

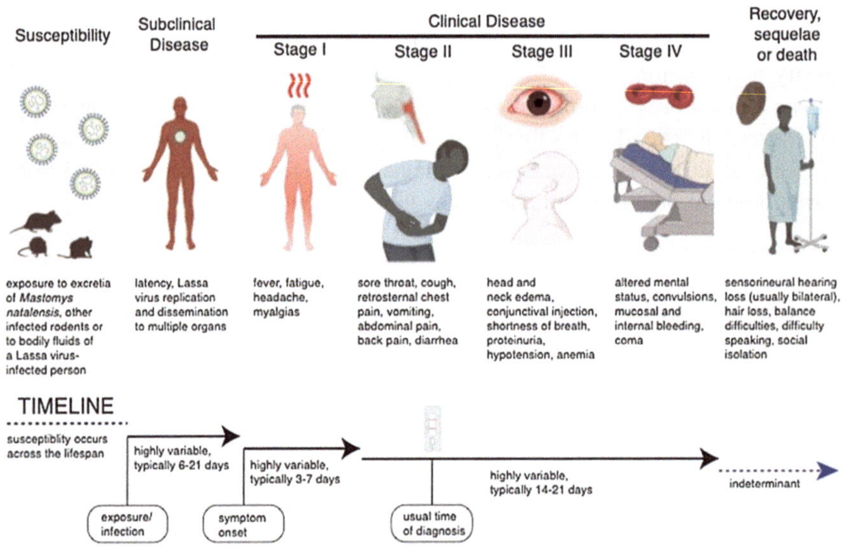

Fig. 1 Natural history of Lassa fever. Lassa fever is a zoonotic infection that has a variable clinical course. After a prolonged but highly variable latency phase, patients may present a spectrum of illnesses. Stages in clinical severity are associated with a variety of signs and symptoms. The major sequelae of Lassa fever typically involve auditory deficits

Leonean children had a temperature of 38 °C or higher at admission (Samuels et al. 2020). Among these antigen-positive children, the most common presenting symptoms were cough (86%), vomiting (75%), and headache (74%), followed by sore throat (58%), pain (58%), head/neck edema (56%), and diarrhea (56%). Unexplained bleeding was reported by 48% of pediatric patients and confusion by 35%. Thus, children generally met major criteria for Lassa fever in adults, including signs and symptoms from previous studies that have been classified as "late presentations." Cases of congenital LASV infection occurring more than 5 days after delivery have also been reported (Monson et al. 1987).

Pathogenic mechanisms of LASV have not been established. Hepatocellular, adrenal, and splenic necrosis are common findings from a limited number of human autopsies combined with non-human primate (NHP) necropsy studies (Baize et al. 2009; Downs et al. 2020; Fisher-Hoch and McCormick 1987; Hensley et al. 2011; Walker et al. 1982; Winn and Walker 1975; Yun and Walker 2012). Nonspecific histopathologic changes, including hemorrhage, edema, and inflammation, occur in the liver, adrenals, and spleen, as well as in the heart, lung, and kidney. The extent of histopathologic changes does not appear sufficient to explain the clinical severity of LASV infection (Shieh et al. 2021). LASV antigens and viral particles have been observed in multiple organs and cells (Hensley et al. 2011; Shieh et al. 2021; Stein et al. 2021). Immunolocalization of LASV antigens in autopsy material from Sierra Leonean patients found extensive involvement of the mononuclear phagocytic system, including tissue macrophages and endothelial cells; suggesting that inflammatory mediators may contribute to vascular dilation and increased permeability during Lassa fever (Shieh et al. 2021).

3 Lassa Fever Patient Demographics

The Viral Hemorrhagic Fever Consortium has worked with the Kenema Government Hospital (KGH), a major center for the study of Lassa fever located in the Eastern Province of Sierra Leone since 2004 (Khan et al. 2008). Other important centers for Lassa fever basic and clinical research are located at the National Public Health Institute of Liberia (NPHIL) (Jetoh et al. 2022), and in Nigeria the Irrua Specialist Teaching Hospital (ISTH) (Akpede et al. 2018b; Ehichioya et al. 2012; Okokhere et al. 2018a), and Federal Medical Centres at Owo and Abakaliki (Saleh et al. 2022). The African Center of Excellence for the Genetics of Infectious Diseases (ACEGID) has conducted sequencing of emerging viruses including LASV and is housed at Redeemers University located in Ede (Happi et al. 2021; Oguzie et al. 2022; Tessema et al. 2020).

Clinical studies in Sierra Leone and Nigeria have revealed differences in the demographics and clinical presentations of patients in these two countries (Table 1) (Ilori et al. 2019; McCormick et al. 1987; Okokhere et al. 2018a; Schieffelin et al. 2023). Patients in Sierra Leone were older than those in Nigeria. There were consistently more female patients than males in Sierra Leone, whereas it is vice versa in Nigeria. Case Fatality Rates (CFRs) were also distinctly different. Because the McCormick study included subjects based solely on a positive antibody test, it is likely that convalescent patients were included in that study accounting for the lower CFR. Signs and symptoms associated with Lassa fever were also distinct across the studies (Table 2). While fever, headache, vomiting, and diarrhea were common across the four studies, sore throat, cough, conjunctivitis, and bleeding were more common in Sierra Leonean patients than in Nigerian patients.

4 Case Identification and Diagnosis

A standardized case definition for Lassa fever developed by the World Health Organization and modified by staff at KGH and Tulane is used to determine admission to the Viral Hemorrhagic Fever Ward, along with, as always, the clinician's judgment (Table 3). It defines common symptoms such as gradual onset of fever, malaise, and general weakness; after a few days, extending to headache, sore throat, muscle pain, chest pain, nausea, vomiting, diarrhea, cough, and abdominal pain; as well as possible bleeding, neck/facial swelling, and shock in severe cases. Similar suspected case definitions have been developed and utilized by the Nigeria Centre for Disease Control (NCDC).

Table 1 Demographics of Lassa fever patients

	McCormick et al. (1987)	Schieffelin et al. (2023)	Okokhere et al. (2018a, b)	Ilori et al. (2019)
Patient location	Sierra Leone	Sierra Leone	Nigeria	Nigeria
Collection period	February 1977–January 1979	2010–2019	January 3, 2011–December 11, 2015	January 1–May 6, 2019
Laboratory confirmation	Virus isolation, IFA titer**, IgG titer $\geq$ 1:256 plus IgM titer $\geq$ 1:4	Antigen ELISA	RT-PCR	IgM ELISA, RT-PCR, virus isolation
Confirmed cases	306***	233*	291	423 (414 lab confirmed)
Median age in years (IQR)	N/A	20 (6–30)	35****	32 (20–44) (n = 414)
Female (%)	56.5	55.6	41.6	37.9
Fatal outcome (%)	16.5	69.5	24	25.1

*Limited to cases with known outcome; **IFA titer seroconversion defined as antibody titer rising from < 1:4 to $\geq$ 1:16; ***Limited to patients 11 years of age and older; symptomatology available for 306 of 441 confirmed patients; ****Mean (Median age was not presented)

Table 2 Proportion of confirmed Lassa fever cases presenting signs and symptoms

	McCormick et al. (1987)*, n = 306	Schieffelin et al. (2023), n = 223	Okokhere et al. (2018a, b), n = 291	Ilori et al. (2019), n = 423
Patient location	Sierra Leone	Sierra Leone	Nigeria	Nigeria
Fever (%)	97	89	ND	96
Weakness/ fatigue (%)	80	79	56	43
Malaise (%)	60	83	ND	ND
Retrosternal chest pain (%)	72	38	24	21
Head and neck edema (%)	ND	35	12	ND
Headache (%)	55	82.8	55	59
Dizziness (%)	60	57	21	ND
Vomiting (%)	48	70	63	49
Sore throat (%)	59	59	38	22
Abdominal pain (%)	49	51	53	40
Cough (%)	55	76	31	30

(continued)

Table 2 (continued)

	McCormick et al. (1987)*, $n = 306$	Schieffelin et al. (2023), $n = 223$	Okokhere et al. (2018a, b), $n = 291$	Ilori et al. (2019), $n = 423$
Joint pain/ arthralgias (%)	42	17	ND	17
Diarrhea (%)	23	42	30	27
Conjunctivitis (%)	25	28	12	5
Myalgias (%)	23	34	ND	19
Bleeding (%)	10	43	25	17
Severe CNS features*** (%)	ND	30	16	ND
Jaundice (%)	ND	8	4	2
Proteinuria (%)	44	ND	66	ND

*Estimated from figure; **Non-severe CNS features "dizziness, lethargy, and drowsiness"; ***Severe central nervous system features "including coma or seizure (irrational talk or behavior, altered sensorium, tremors, and disorientation or confusion which suggest encephalitis, meningitis, or encephalopathy)"
For overlapping categories (e.g., convulsions and altered mental status) the category with the higher proportion is presented. Signs and symptoms were only included in the table if data was present in 2 or more studies. McCormick et al. (1987) described 50% with pharyngitis and 55% with back pain; Okokhere et al. (2018a, b) described 21% with non-severe CNS features and 61% with hematuria; Ilori et al. (2019) described 15% with dyspnea, 33% with anorexia, 4% with disorientation, 5% with unconscious, 3% with photophobia, 3% with hiccups, 4% with rash, and 17% with arthralgias. ND = not described

Table 3 Case definition for suspected cases of Lassa fever

- Known exposure to a person suspected to have Lassa fever ***OR***
- Fever > 38 °C for less than three weeks ***PLUS***
- Absence of signs of local inflammation ***AND***
- Two major signs or one major and two minor signs

Major signs	Minor signs
• Bleeding • Swollen neck or face • Conjunctival injection or sub-conjunctival hemorrhage • Spontaneous abortion • Petechial or hemorrhagic rash • New onset of tinnitus or altered hearing • Persistent hypotension • Absence of clinical response after 48 h to anti-malarial and/or broad-spectrum antibiotic therapy • Elevated liver transaminases, especially AST > ALT	• Headache • Sore throat • Vomiting • Diffuse abdominal pain/ tenderness • Chest/retrosternal pain • Cough • Diarrhea • Generalized myalgia or arthralgia • Profuse weakness • Proteinuria • Leucopenia < 4000/μl

Modified from (Khan et al. 2008)

5 Diagnostic Testing

Prompt diagnosis of LASV infection is imperative for proper clinical management and isolation of patients. However, early diagnosis based on clinical presentation alone is challenging because the initial symptoms, including fever, headache, malaise, and general weakness, are nondescript (McCormick et al. 1987; Mertens et al. 1973; Okokhere et al. 2018a; Shaffer et al. 2014). These symptoms are common to other illnesses such as malaria, typhoid fever, and leptospirosis that frequently occur in Lassa endemic areas, often delaying diagnosis (Ashcroft et al. 2022; Boisen et al. 2015). Laboratory diagnosis of acute Lassa fever in West Africa relies on the detection of LASV proteins or antigens (Ag) using immunoassays or LASV genomic RNA using molecular diagnostics, most commonly polymerase chain reaction (PCR) (Boisen et al. 2018, 2020; Hartnett et al. 2015; Olschlager and Gunther 2012). The diagnosis of Lassa fever by immunological and molecular methods is challenged by the high genetic diversity of LASV (Andersen et al. 2015; Heinrich et al. 2020). The field performance of a first-generation rapid diagnostic test (RDT) versus quantitative PCRs (qPCRs) at KGH showed 90% sensitivity and 100% specificity, better than the most robust qPCR currently available (82% sensitivity and 95% specificity) (Boisen et al. 2018). Longitudinal sampling demonstrates that viral yield is highest early in infection and declines during the second week, which may affect both antigen testing and nucleic acid-based tests. Diagnostic assays using CRISPR-Cas13a have been developed for LASV infection (Barnes et al. 2020).

In addition to Ag and nucleic acid detection, LASV-specific immunoglobulin M (IgM) or immunoglobulin G (IgG) immunoassays can also contribute to the clinical management of Lassa fever patients (Branco et al. 2011b; Heinrich et al. 2020; Shaffer et al. 2021). Serological assays typically yield negative results in the first week of illness before the development of anti-LASV antibodies. There is a transition from LASV antigenemia to IgM and IgG seropositivity (Bausch et al. 2000; Branco et al. 2011a, b; Grove et al. 2011; Ibekwe et al. 2012). Inverse correlations of LASV Ag by RDT or qPCR signal versus IgM and IgG titers have been observed (Boisen et al. 2018, 2020). These results suggest that the presence of a humoral immune response to LASV clears viral antigen and RNA, but that the clearance requires an extended time-period and can be incomplete. Post-acute Lassa fever patients at presentation can include detectable levels of viral nucleic acid, circulating antigen, IgM, IgG, or a combination thereof.

A proposed diagnostic algorithm suggested that suspected LF cases can be triaged using a second-generation Lassa RDT (Boisen et al. 2020). Presumptive results of the RDT would be confirmed by qPCR using pre-established testing cut-offs along with IgM and IgG enzyme-linked immunosorbent assay (ELISA) screening to designate patients as Acute Lassa fever, Post-Acute Lassa fever (Ag−/IgM+), or Non-Lassa fever (including Lassa fever Convalescent patients that are IgG seropositive). This screening algorithm has the potential to reduce false positive results. Decisions can begin to be made in the development time of the RDT, as little as 5–25 min.

6 Other Clinical and Laboratory Testing

Potentially, a broad array of clinical laboratory tests using standard methods can aid in the clinical management of Lassa fever. Thrombocytopenia and transient lymphopenia can occur, along with a monocytosis during the second week of infection (Fisher-Hoch et al. 1988). Electrolyte imbalances, acute renal failure, and acute hepatitis have been well described (Grant et al. 2023a; Okokhere et al. 2018a; Schieffelin et al. 2023). Therefore, serum blood chemistries (sodium, potassium, chloride, carbon dioxide, glucose, calcium, albumin, and total protein), renal function tests (blood urea nitrogen and creatinine), and liver function tests (alanine aminotransferase [ALT], aspartate aminotransferase [AST], amylase, alkaline phosphatase, and total bilirubin) can help support the diagnosis of acute Lassa fever and help guide supportive care measures (Schieffelin et al. 2023; Shaffer et al. 2014). Arterial blood gases are helpful in resource-rich settings.

The underdeveloped settings, where Lassa fever is typically found, often preclude the use of advanced ancillary testing. Clinical management in advanced settings can include chest and abdominal imaging, audiometry, echocardiography, and fetal Doppler assessment in pregnant women. If available, these tests should be performed at admission and then thereafter as the patient's clinical status warrants. The infrastructure provided by advanced facilities in some endemic countries and newer technology makes some equipment more affordable and offers new opportunities for understanding Lassa fever pathogenesis. Pleural and pericardial effusion can occur in Lassa fever (Cummins et al. 1989a; Edington and White 1972; Fisher-Hoch et al. 1987; McCormick et al. 1987; Monath et al. 1974). Assessments of the hemodynamic status and fluids shifts as some patients may descend into shock, including continuous non-invasive hemodynamic monitoring, pulse oximetry, and periodic echocardiography can be of value for clinical assessments. The use of remote sensors (wearables) to collect data allows the capture of physiological attributes continuously throughout the course of LASV infection is also being explored. The use of such devices is particularly of interest in the practical management of patients with Lassa fever and potentially other highly transmissible infectious diseases in settings where other advanced monitoring capacities are limited.

Given the diversity of febrile illnesses in West Africa, it is desirable to attempt to achieve the diagnosis for patients who test negative for Lassa fever or examine the possibility of co-infections with LASV (Roth et al. 2015). Thin and thick smears for malaria or a malaria rapid test and a rapid test for typhoid fever should be performed. If malaria cannot be ruled out in a timely manner, empiric treatment is warranted, especially in children under five years of age. Other hemorrhagic fevers, including Ebola, have emerged in West Africa (Boisen et al. 2016; Goba et al. 2016). Ruling out Ebola, particularly in communities that previously had large numbers of cases, is prudent. In selected cases, and with specific pre- and post-test counseling and consent, testing for HIV may also be performed.

7 Prognostic Indicators

CFRs of Lassa fever patients that present for clinical care vary widely, ranging from 24 to 60% in Nigeria to up to 69% in Sierra Leone (Buba et al. 2018; Grant et al. 2023a; Ilori et al. 2019; McCormick et al. 1987; Okokhere et al. 2018a; Schieffelin et al. 2023; Shaffer et al. 2014; 2019, 2021). Factors hypothesized to affect outcome include variation in LASV strain, patient characteristics, and time from symptom onset to presentation as well as disease severity at the time of presentation. LASV strains demonstrate significant genetic diversity, with variations of up to 32% at a nucleotide level in the L RNA strand and 25% in the S RNA strand (Andersen et al. 2015; Bowen et al. 2000) across as many as 8 distinct genetic lineages. Viral genetic diversity may account in part for the variability in clinical presentation.

An increased risk of death from Lassa fever has been associated with several patient characteristics, including pregnancy and high virus load (Branco et al. 2011b; Grove et al. 2011; Kayem et al. 2020; Shaffer et al. 2014, 2021). Pregnant women are consistently recognized as having increased odds of death after the development of Lassa fever. In pregnant patients with Lassa fever, spontaneous abortion or fetal demise before or after birth occurs commonly (Branco et al. 2011a; Kayem et al. 2020; Price et al. 1988; Walls 1985). An overall odds ratio of 2.86 for death compared to non-pregnant women with LF and an odds ratio of over 5 when presenting in the third trimester have been estimated (Kayem et al. 2020). In a study, 30 pregnant Nigerian Lassa fever patients were treated with early ribavirin therapy and a conservative obstetric approach at a teaching hospital in southern Nigeria. It was found that, among them, 11/30 (36.7%) women died, and 20/31 (64.5%) pregnancies ended in fetal or perinatal loss. On initial evaluation, 17/30 (56.6%) women had a dead fetus; 10/17 (58.8%) of these patients died, compared to 1/13 (7.7%) of women with a live fetus (Okogbenin et al. 2019). Extravaginal bleeding, convulsions, and oliguria were independently associated with maternal and fetal or perinatal death. While generally consistent with previous observations, the positive outcomes in pregnant patients with a living fetus may support a more conservative approach to obstetric management of Lassa fever in pregnancy in Nigeria. Among Sierra Leonean women of childbearing age, pregnant Ag−/IgM + women were more likely to die, although this difference was not seen among Ag + women (Grant et al. 2023a). In both groups of Sierra Leonean women, fatalities were more likely to be present with altered mental status.

A number of other markers have been correlated with poor outcomes of Lassa fever, including elevated serum AST and the presence of acute kidney injury (Kidney Disease-Improving Global Outcome grade $\geq$ 2; stage $\geq$ 2 acute kidney injury in children) (Duvignaud et al. 2020; Grant et al. 2023a; Schieffelin et al. 2023; Shaffer et al. 2014; Yasri and Wiwanitkit 2018). Individual patient characteristics associated with death across studies have often yielded inconsistent findings. These discrepancies can only partly be explained by the data collected, leaving uncertainty over the clinical and laboratory associations with a fatal outcome among LF patients. At KGH, we examined acute LF patients with detectable LASV antigen (Ag+) as well as patients with antigen negative/IgM positive (Ag−/IgM+) results, which we have referred to as subacute Lassa fever and antigen negative/immunoglobulin negative febrile controls (Schieffelin et al. 2023). Both the Ag+ and Ag−/IgM+ patient groups had high but significantly different CFRs. In our cohort of 233 Ag+ and 243 Ag−/IgM+ Lassa fever cases, the CFR was 69.5% and 25.1%, respectively. Several differences were also noted in laboratory parameters. Ag+ fatal cases were more likely to present with hypokalemia, acidosis, and evidence of renal dysfunction and hypoalbuminemia. Ag−/IgM+ fatal cases only demonstrated an increased frequency of capillary leak compared to Ag−/IgM+ survivors. While elevated AST was associated with death in previous studies, we only found that hyperbilirubinemia and an elevated AST to ALT ratio were significant. This inconsistency is likely due to the small number of AST results available in our sample. A more advanced stage of disease at the time of presentation, irrespective of time to presentation, was associated with an increased risk of death among Ag+ cases.

Elevated clinical scores, such as the National Early Warning Score (Version 2), have been shown to be predictive of a fatal outcome when elevated ($\geq$ 7) (Chiosi et al. 2022; Grant et al. 2023a). Three existing clinical prediction tools have been evaluated on their ability to predict the in-hospital mortality in Lassa fever: the quick Sequential Organ Failure Assessment (qSOFA), the Modified Early Warning System (MEWS), and the Universal Vital Assessment (UVA) (Chiosi et al. 2022). In a retrospective cohort study of patients admitted to the dedicated Lassa fever ward of the Kenema Government Hospital in Sierra Leone between May 2013 and December 2019, data among three serology groups were analyzed: Lassa antigen-positive (Ag+) regardless of IgM status, Lassa Ag− and IgM+, and Lassa Ag− and IgM− cases. There were 123 cases of suspected Lassa fever included in this study. Abnormalities in respiratory rate, oxygenation status, mental status, and serum markers of kidney and liver dysfunction were more likely seen in the Ag+ group, which had an in-hospital mortality of 85.7%. For the Lassa Ag+ group, the sensitivity and positive predictive value of qSOFA $\geq$ 2 was 70.6 and 92.3%, MEWS $\geq$ 5 was 96.9 and 86.1%, and UVA $\geq$ 5 was 60.0 and 100.0%. The MEWS and UVA scores show potential for use in Lassa fever, but there is an opportunity for future development of a tool that includes the clinical and laboratory markers specific to Lassa fever.

Pre-existing immunity to LASV may play a protective role as it is rare for patients with acute Lassa fever to present with IgG responses to the virus (Shaffer et al. 2014, 2021). However, only limited clinical data from Lassa fever patients is available to assess the role of immune responses in the diverse outcomes that occur following LASV infection. LASV suppresses various aspects of innate immunity (Meyer and Ly 2016; Shao et al. 2015). For example, the nucleoprotein possesses a double-stranded RNA exonuclease activity capable of blunting the interferon response (Hastie et al. 2011). Humoral immune responses do not appear to play a major role in determining Lassa fever outcome. Most Lassa fever survivors do not produce neutralizing antibodies until weeks or months into recovery (Robinson et al. 2016). In contrast, cellular immunity does appear to play an important role in surviving Lassa fever (Baize et al. 2009; Flatz et al. 2010; Hayes et al. 2012; LaVergne et al. 2022; Lukashevich et al. 2019; Sakabe et al. 2020; Sullivan et al. 2020). Examination of patients during the 2018 Lassa fever Nigerian outbreak in Nigeria found that severe Lassa fever is associated with poor LASV-specific effector T-cell responses (Port et al. 2020). Immunopathogenesis may also be involved. Bystander T cells that respond to antigen unrelated to LASV were also activated in patients with severe Lassa fever. These cells expressed defined tissue homing signatures that suggest their recruitment to inflamed tissues.

Children may have a different set of prognostic indicators compared to adults (Samuels et al. 2020; Webb et al. 1986). Hospitalized children with Lassa fever in Sierra Leone with positive antigen tests experience high mortality (63%) (Samuels et al. 2020). Confusion, unexplained bleeding, and head/neck edema were common among our cohort of children with LF antigen–positive results who died. Univariate analyses showed a statistically significant higher OR of mortality in children with confusion and unexplained bleeding.

Further clinical studies may reveal factors associated with fatal outcomes. Patients with severe Lassa fever can exhibit abnormal coagulation, endothelial barrier disruption, and dysfunctional platelet aggregation (Cummins et al. 1989b; Fisher-Hoch et al. 1988; Horton et al. 2020; Zapata et al. 2014). On the day-of-admission, plasma samples from Lassa fever patients in Sierra Leone were tested for levels of proteins necessary for coagulation, fibrinolysis, and platelet function (Horton et al. 2020). P-selectin, soluble endothelial protein C receptor, soluble thrombomodulin, plasminogen activator inhibitor 1, ADAMTS-13, von Willebrand factor, tissue factor, soluble intercellular adhesion molecule 1, and vascular cell adhesion molecule 1 were more elevated in Lassa fever patients than in controls. Endothelial protein C receptor, thrombomodulin, intercellular adhesion molecule 1, plasminogen activator inhibitor 1, D-dimer, and hepatocyte growth factor were higher in fatal than nonfatal Lassa fever cases. Platelet disaggregation occurred only in samples from fatal Lassa fever cases. The impaired homeostasis and platelet dysfunction implicate alterations in the protein C pathway, which might contribute to the loss of endothelial barrier function in fatal infections.

Tools of systems biology may provide insight into factors that can predict a fatal outcome in Lassa fever (Gale et al. 2017, 2020). Metabolomics was used to identify pathways that are differentially affected during Lassa fever and guide the development of diagnostics to monitor and predict the outcome. Effects of LASV infection on pathways mediating blood coagulation, and lipid, amino acid, and nucleic acid metabolism manifest in changes in the levels of numerous metabolites in the circulation. Several compounds, including platelet-activating factor (PAF), PAF-like molecules, and products of heme breakdown emerged as candidates that may prove useful in diagnostic assays to inform better care of Lassa fever patients. The amino acid L-threonine is elevated during LASV infection.

8 Treatment

The mainstay of Lassa fever treatment is supportive care, particularly maintenance of oxygenation, electrolyte balance, and blood pressure (WHO 2016). Secondary damage to kidneys, lungs, and other organs or exacerbation of co-morbidities often occur during Lassa fever and must be managed clinically. Fluid replacement is important, although volume overload due to the potential capillary leak and renal dysfunction must be monitored and avoided. Aspirin and nonsteroidal anti-inflammatory drugs should not be used because they can adversely affect blood coagulation and clotting thereby accelerating hemorrhage. Because of the high incidence of acute kidney injury among Lassa fever patients, access to renal replacement therapy likely provides a survival benefit (Adetunji et al. 2021; Olatunde et al. 2021). However, guidelines on standard of care supportive therapy, while available, are based on limited clinical evidence and occasionally contradictory (Rigby et al. 2023).

Although no drug has been approved, ribavirin [1-(β-D-Ribofuranosyl)-1″H″-1,2,4-triazole-3-carboxamide], in conjunction with supportive care, is currently the primary treatment for Lassa fever (Groger et al. 2023; Hartnett et al. 2015; Melnik 2022; Salam et al. 2022). Early studies suggested that ribavirin administration within the first 6 days after the onset of fever can reduce case mortality to as low as 5% (Dahmane et al. 2014; McCormick et al. 1986). However, these studies may have overestimated the efficacy of ribavirin due to the inclusion of convalescent or other post-acute subjects (Salam et al. 2021). Patients at the post-acute stage of LASV infection whose immune response has controlled the LASV load, yet remain sick enough to seek treatment, may not benefit from ribavirin treatment. These later-stage patients may, however, benefit from supportive therapies such as fluid replacement.

Clinical trials have not yet confirmed the overall efficacy or safety of ribavirin for the treatment of Lassa fever (Cheng et al. 2022; Salam et al. 2022). A recent meta-analysis suggested that ribavirin reduces mortality for individuals with elevated AST but is potentially harmful in those without elevated AST (Eberhardt et al. 2019). Some experts withhold ribavirin for individual patients with mild to no symptoms and an aminotransferase level (AST) < 150 international units (IU)/L. Ribavirin carries a risk of hemolytic anemia or rigors (Fisher-Hoch et al. 1992). It is also teratogenic and embryotoxic presenting risks of treatment for pregnant and breastfeeding women (Ferm et al. 1978). Changes in mood, dizziness, fatigue, fever, and other side effects are common with ribavirin treatment.

There are few studies to date on pediatric Lassa fever making it difficult to fully understand best practices for management. A retrospective study of pediatric patients with LF in Sierra Leone did not show a significant survival benefit associated with ribavirin treatment (Samuels et al. 2020). Among the 57 children with LASV antigen-positive results, 38 (67%) received ribavirin, of which 23 (64%) died. Receipt of ribavirin was not significantly associated with survival among these LASV antigen-positive children.

Favipiravir (6-fluoro-3-hydroxy-2-pyrazine carboxamide) is an oral viral RNA polymerase inhibitor that has been licensed for the treatment of influenza in Japan (Hansen et al. 2021; Melnik 2022). Treatment improved survival from LASV challenge in various animal models including NHPs (Mendenhall et al. 2011; Safronetz et al. 2015). A ribavirin and favipiravir combination was used in 2 individuals in an advanced clinical setting, both of whom survived (Raabe et al. 2017).

Convalescent plasma has shown mixed benefits in improving survival from LASV infection in animal models (Frame et al. 1984; Jahrling and Peters 1984; Jahrling et al. 1984; Leifer et al. 1970; McCormick et al. 1986). Success appeared to be dependent on the match of neutralizing antibodies to the LASV strain causing infection (Jahrling et al. 1985; Jahrling and Peters 1984). Recently recovered individuals may not be suitable for convalescent plasma donation because neutralizing antibodies against LASV often take several months to develop, if at all, after natural infection (Robinson et al. 2016). Non-human primate data suggested that a combination of ribavirin therapy with convalescent plasma may be more effective at preventing fatal Lassa fever compared to either administered alone (Jahrling et al. 1984). However, limited availability and the challenge of maintaining lot-to-lot consistency likely limits the widespread use of convalescent plasma. Arevirumab-3, a combination of 3 neutralizing human monoclonal antibodies that target different epitopes on the LASV glycoprotein, protects against fatal Lassa fever in NHPs, even when given a week after infection (Cross et al. 2019, 2016; Li et al. 2022; Mire et al. 2017). Preclinical development of the antibody cocktail is nearly completed, and Arevirumab-3 will soon enter human clinical trials.

9 Sequelae

In studies from Nigeria, Lassa fever survivors with severe acute kidney injury, including those who required hemodialysis, recovered renal function to normal levels (Okokhere et al. 2018a). However, long-term sequela has been reported following LASV infection. The most reported long-term consequence of Lassa fever is bilateral sensorineural hearing loss (Ficenec et al. 2019; Ibekwe et al. 2011; Mateer et al. 2018; Simmons 1993). Audiometry results reveal that Lassa fever survivors in Sierra Leone had significantly more bilateral hearing loss in comparison to matched controls from the community (30% vs. 4%, $p = 0.029$) (Ficenec et al. 2020). Lassa fever survivors were significantly more likely to report balance difficulties (55% vs. 20%, $p < 0.001$), hair loss (32% vs. 7%, $p < 0.001$), difficulty speaking (19% vs. 1%, $p < 0.001$), social isolation (50% vs. 0%, $p < 0.001$), and hearing loss (17% vs. 1%, $p = 0.002$) in comparison to controls. Hearing loss may involve damage to cochlear hair cells or the auditory nerve due to perivascular inflammation (Gary et al. 2019; Huynh et al. 2020; Yun et al. 2015). Evidence from Lassa virus-infected NHPs directly implicated immune-mediated vasculitis in the pathology of Lassa fever-associated deafness (Cashman et al. 2018). Social isolation and stigmatization, with or without hearing loss, remain a serious challenge to the mental health of Lassa fever survivors (Ficenec et al. 2020; Usifoh et al. 2019). Lassa fever survivors in a cohort in Sierra Leone were also found to have cataract and posterior segment findings that potentially represent sequelae of uveitis associated with visual impairment similar to findings among survivors of the Ebola virus disease (Li et al. 2020).

10 Post-exposure Prophylaxis

Post-exposure prophylaxis maybe warranted for contacts of known or suspected Lassa fever patients that have risk factors for LASV transmission (Bausch et al. 2010; Hadi et al. 2010). Such potential exposures include penetrating needle stick injury, broken skin or mucous membrane contact with patient blood or body fluids, or participation in clinical procedures without personal protective equipment (PPE). The recommended protocol for Lassa prophylaxis is a 35 mg/kg loading dose of ribavirin followed by 15 mg/kg orally (maximum dose 1 g) every eight hours for 10 days. The effectiveness of this regimen has not been evaluated in a controlled clinical trial. Regardless of whether prophylactic ribavirin is prescribed, close monitoring and diagnostic testing of high-risk contacts are recommended for 21 days. Those who develop fever (temperature $\geq$ 38.3 °C) should begin treatment with intravenous ribavirin.

Risks versus benefits of post-exposure prophylaxis are important to be considered based on known side effects and toxicities of ribavirin. Favipiravir, could potentially be a useful substitute as a prophylactic option, but like ribavirin, should be avoided in pregnant and breastfeeding women because of teratogenicity, embryotoxicity, and potential excretion in breast milk (Nagata et al. 2015). Studies with Arevirumab-3 in NHPs suggest that it would be an effective prophylactic option in the event of potential LASV exposure (Cross et al. 2019, 2016; Mire et al. 2017). Like other antiviral monoclonal antibodies, Arevirumab-3 has not been associated with side effects in animal models.

11 Infection Prevention and Control

Facilities that manage Lassa fever patients must be carefully designed and access highly regulated (WHO 2016). Only health care personnel trained in the use of appropriate PPE should attend to patients. Areas, where patients or infectious material is present, should be accessible via controlled pathways. The VHF Ward at KGH provides an example of how entry to and exit from patient rooms can be managed (Fig. 2). Patients, particularly during the acute phase of illness, should be placed under high containment in a single-patient room with a private bathroom.

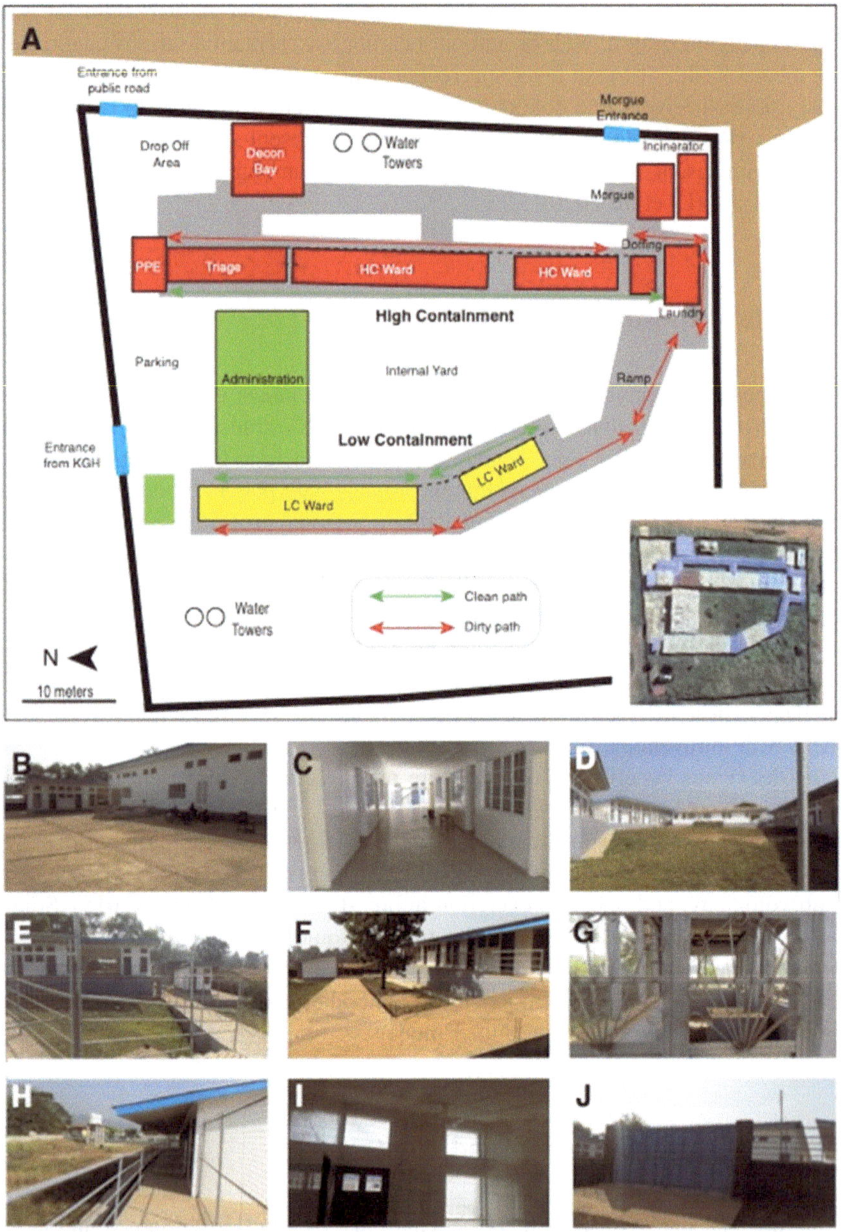

Fig. 2 The Viral Hemorrhagic Fever Ward at Kenema Government Hospital. The VHF Ward is designed to isolate Lassa fever patients and other patients with highly transmissible diseases into low containment (LC) and high containment (HC) wards. Panel A: Plan of the VHF Ward. Inset: Satellite view using Google Earth. Panel B: View of the Administration building from the car park. Panel C: Interior corridor of the administration building. Panel D: View from the internal yard. Panel E: View from the ramp showing the doffing room and laundry. Panel F: Drop-off area. Panel G: Incinerator. Panel H: Clean path of the low containment ward. Panel I: Room interior. Panel J: Ambulance entrance from road into drop-off area. Photographs taken prior to the addition of covered walkways

PPE used for Lassa patient management should minimally include 2 pairs of gloves, a fluid-resistant, disposable gown or coverall, a waterproof apron, head and neck coverings, eye protection, and waterproof boots/shoe covers. An N95 respirator is preferred for all patient care activities but should always be used during procedures that generate aerosols, such as childbirth. Autopsies should be performed with extreme care because of the high transmission risk. Donning and doffing protocols for PPE should be established and frequent hand hygiene is important. The use of sharps should be minimized, and disposal units should be accessible. Liquid waste management needs may be high as Lassa patients can produce high volumes of stool.

Appropriate PPE should also be used by laboratory workers performing diagnostic assays on samples from patients with suspected LASV infections. Biosafety class II cabinets should be used to perform laboratory assays. In both clinical and laboratory settings, plans should be developed for the appropriate decontamination of equipment, including ambulances, after use.

12 Interventions at the Community Level

Detection of a case of Lassa fever should activate a public health response in the community. Active and passive surveillance for additional cases of Lassa fever should be conducted in collaboration with the Ministries of Health and local health centers. Teams should be well versed in survey methods including sampling strategies, questionnaire administration, and collection of blood and other specimens in community settings, and have extensive knowledge of the communities. Ideally, community outreach teams include members who speak languages spoken in the affected areas. Suspected cases of Lassa fever should be identified, and contacts should be traced in the community. Measures that should be implemented in the community include the avoidance of rodents and the isolation of potentially infected individuals. In local healthcare facilities, enhanced precautions (standard, contact, and droplet) should be implemented, and all blood and bodily fluids should be handled with caution.

Community education for the prevention and early detection of VHF, including implementation of Lassa fever rodent control interventions and sanitation, should be maintained for at-risk communities. Experienced rodent control and hygiene can provide a valuable resource for the development of Lassa fever prevention programs. Expertise in field survey methods, including live-capture trapping and sample preservation and storage, is an asset for controlling Lassa fever at the community level.

13 Summary

More than fifty years after the initial clinical description and discovery of the etiological agent, Lassa fever remains a persistent public health threat in West Africa. Morbidity and mortality caused by LASV infection remain high, especially in pregnant women. Various options are now available to aid with Lassa fever diagnosis. However, they are generally not available in rural clinics where most patients present but are restricted to major research centers. The pathophysiology of Lassa fever in humans is still poorly understood. Supportive care, including fluid resuscitation, remains the mainstay of Lassa fever treatment and management with as yet no specific therapeutics available that have received regulatory approval. Early initiation of intravenous ribavirin may confer some survival benefits; however, this drug has numerous side effects and is teratogenic. Favipiravir shows potential benefits, but it is an oral drug that is difficult to administer to extremely ill patients at high frequent doses. A new monoclonal antibody therapy for LASV infection is in development, which shows promising results in NHPs for treatment and perhaps prophylaxis. Well-conducted randomized trials of current and new therapeutics are needed. No Lassa fever vaccines are currently approved for use although several promising vaccine candidates are in the early stages of human clinical trials. Community education on the role of rodents in Lassa fever transmission and the implementation of appropriate control procedures should remain a public health priority.

References

Abudurexiti A, Adkins S, Alioto D, Alkhovsky SV, Avsic-Zupanc T, Ballinger MJ, Bente DA, Beer M, Bergeron E, Blair CD, Briese T, Buchmeier MJ, Burt FJ, Calisher CH, Chang C, Charrel RN, Choi IR, Clegg JCS, de la Torre JC, de Lamballerie X, Deng F, Di Serio F, Digiaro M, Drebot MA, Duan X, Ebihara H, Elbeaino T, Ergunay K, Fulhorst CF, Garrison AR, Gao GF, Gonzalez JJ, Groschup MH, Gunther S, Haenni AL, Hall RA, Hepojoki J, Hewson R, Hu Z, Hughes HR, Jonson MG, Junglen S, Klempa B, Klingstrom J, Kou C, Laenen L, Lambert AJ, Langevin SA, Liu D, Lukashevich IS, Luo T, Lu C, Maes P, de Souza WM, Marklewitz M, Martelli GP, Matsuno K, Mielke-Ehret N, Minutolo M, Mirazimi A, Moming A, Muhlbach HP, Naidu R, Navarro B, Nunes MRT, Palacios G, Papa A, Pauvolid-Correa A, Paweska JT, Qiao J, Radoshitzky SR, Resende RO, Romanowski V, Sall AA, Salvato MS, Sasaya T, Shen S, Shi X, Shirako Y, Simmonds P, Sironi M, Song JW, Spengler JR, Stenglein MD, Su Z, Sun S, Tang S, Turina M, Wang B, Wang C, Wang H, Wang J, Wei T, Whitfield AE, Zerbini FM, Zhang J, Zhang L, Zhang Y, Zhang YZ, Zhang Y, Zhou X, Zhu L, Kuhn JH (2019) Taxonomy of the order Bunyavirales: update 2019. Arch Virol 164:1949–1965

Adetunji AE, Ayenale M, Akhigbe I, Akerele LO, Isibor E, Idialu J, Aideloje FO, Emuebonam E, Aire C, Adomeh DI, Odia I, Atafo RO, Okonofua MO, Owobu A, Ogbaini-Emovon E, Tobin EA, Asogun DA, Okogbenin SA, Sabeti P, Happi CT, Günther S, Azubuike CO, Rafiu M, Odike A, Olomu SC, Ibadin MO, Okokhere PO, Akpede GOP (2021) Acute kidney injury and mortality in pediatric Lassa fever versus question of access to dialysis. Int J Infect Dis IJID Official Publ Int Soc Infect Dis 103:124–131

Akhiwu HO, Yiltok ES, Ebonyi AO, Gomerep S, Shehu NY, Amaechi EP, Onukak AE, Iduh P, Oyagbemi AA, Omame G, Ashir PM, Egah DZ, Oguche S (2018) Lassa fever outbreak in adolescents in North Central Nigeria: report of cases. J Virus Erad 4:225–227

Akpede GO, Adetunji AE, Udefiagbon EO, Eluehike SO, Odike AI, Ewah-Odiase RO, Omosofe FO, Akerele JM, Atafo RO, Okonofua MO, Onyeke I, Oyedeji OA, Mbanefo K, Ogbaini-Emovon E, Okokhere PO, Gunther S, Dongo AE (2018a) Acute abdomen in pediatric patients with Lassa fever: prevalence and response to nonoperative management. J Pediatr Infect Dis Soc

Akpede GO, Asogun DA, Okogbenin SA, Dawodu SO, Momoh MO, Dongo AE, Ike C, Tobin E, Akpede N, Ogbaini-Emovon E, Adewale AE, Ochei O, Onyeke F, Okonofua MO, Atafo RO, Odia I, Adomeh DI, Odigie G, Ogbeifun C, Muebonam E, Ihekweazu C, Ramharter M, Colubri A, Sarbeti PC, Happi CT, Gunther S, Agbonlahor DE (2019) Caseload and case fatality of Lassa fever in Nigeria, 2001–2018: a specialist center's experience and its implications. Front Public Health 7:170

Akpede GO, Asogun DA, Okogbenin SA, Okokhere PO (2018b) Lassa fever outbreaks in Nigeria. Expert Rev Anti Infect Ther 16:663–666

Andersen KG, Shapiro BJ, Matranga CB, Sealfon R, Lin AE, Moses LM, Folarin OA, Goba A, Odia I, Ehiane PE, Momoh M, England EM, Winnicki S, Branco LM, Gire SK, Phelan E, Tariyal R, Tewhey R, Omoniwa O, Fullah M, Fonnie R, Fonnie M, Kanneh L, Jalloh S, Gbakie M, Saffa S, Karbo K, Gladden AD, Qu J, Stremlau M, Nekoui M, Finucane HK, Tabrizi S, Vitti JJ, Birren B, Fitzgerald M, McCowan C, Ireland A, Berlin AM, Bochicchio J, Tazon-Vega B, Lennon NJ, Ryan EM, Bjornson Z, Milner DA Jr, Lukens AK, Broodie N, Rowland M, Heinrich M, Akdag M, Schieffelin JS, Levy D, Akpan H, Bausch DG, Rubins K, McCormick JB, Lander ES, Gunther S, Hensley L, Okogbenin S, Schaffner SF, Okokhere PO, Khan SH, Grant DS, Akpede GO, Asogun DA, Gnirke A, Levin JZ, Happi CT, Garry RF, Sabeti PC (2015) Clinical sequencing uncovers origins and evolution of Lassa virus. Cell 162:738–750

Ashcroft JW, Olayinka A, Ndodo N, Lewandowski K, Curran MD, Nwafor CD, Madubuike K, Bagnall H, Musa AS, Ahmad A, Akinpelu A, Adeleye A, Martin C, Akano A, Onoja M, Iniobong A, Ukponu W, Arinze C, Ilori E, Mba N, Gannon B, Ahumibe A, Ihekweazu C (2022) Pathogens that cause illness clinically indistinguishable from Lassa fever, Nigeria, 2018. Emerg Infect Dis 28:994–997

Baize S, Marianneau P, Loth P, Reynard S, Journeaux A, Chevallier M, Tordo N, Deubel V, Contamin H (2009) Early and strong immune responses are associated with control of viral replication and recovery in Lassa virus-infected cynomolgus monkeys. J Virol 83:5890–5903

Barnes KG, Lachenauer AE, Nitido A, Siddiqui S, Gross R, Beitzel B, Siddle KJ, Freije CA, Dighero-Kemp B, Mehta SB, Carter A, Uwanibe J, Ajogbasile F, Olumade T, Odia I, Sandi JD, Momoh M, Metsky HC, Boehm CK, Lin AE, Kemball M, Park DJ, Branco L, Boisen M, Sullivan B, Amare MF, Tiamiyu AB, Parker ZF, Iroezindu M, Grant DS, Modjarrad K, Myhrvold C, Garry RF, Palacios G, Hensley LE, Schaffner SF, Happi CT, Colubri A, Sabeti PC (2020) Deployable CRISPR-Cas13a diagnostic tools to detect and report Ebola and Lassa virus cases in real-time. Nat Commun 11:4131

Basinski AJ, Fichet-Calvet E, Sjodin AR, Varrelman TJ, Remien CH, Layman NC, Bird BH, Wolking DJ, Monagin C, Ghersi BM, Barry PA, Jarvis MA, Gessler PE, Nuismer SL (2021) Bridging the gap: using reservoir ecology and human serosurveys to estimate Lassa virus spillover in West Africa. PLoS Comput Biol 17:e1008811

Bausch DG, Hadi CM, Khan SH, Lertora JJ (2010) Review of the literature and proposed guidelines for the use of oral ribavirin as postexposure prophylaxis for Lassa fever. Clin Infect Dis 51:1435–1441

Bausch DG, Rollin PE, Demby AH, Coulibaly M, Kanu J, Conteh AS, Wagoner KD, McMullan LK, Bowen MD, Peters CJ, Ksiazek TG (2000) Diagnosis and clinical virology of Lassa fever as evaluated by enzyme-linked immunosorbent assay, indirect fluorescent-antibody test, and virus isolation. J Clin Microbiol 38:2670–2677

Bell-Kareem AR, Smither AR (2021) Epidemiology of Lassa fever. Curr Top Microbiol Immunol

Boisen ML, Hartnett JN, Goba A, Vandi MA, Grant DS, Schieffelin JS, Garry RF, Branco LM (2016) Epidemiology and management of the 2013–16 West African Ebola Outbreak. Ann Rev Virol

Boisen ML, Hartnett JN, Shaffer JG, Goba A, Momoh M, Sandi JD, Fullah M, Nelson DKS, Bush DJ, Rowland MM, Heinrich ML, Koval AP, Cross RW, Barnes KG, Lachenauer AE, Lin AE, Nekoui M, Kotliar D, Winnicki SM, Siddle KJ, Gbakie M, Fonnie M, Koroma VJ, Kanneh L, Kulakosky PC, Hastie KM, Wilson RB, Andersen KG, Folarin OO, Happi CT, Sabeti PC, Geisbert TW, Saphire EO, Khan SH, Grant DS, Schieffelin JS, Branco LM, Garry RF (2018) Field validation of recombinant antigen immunoassays for diagnosis of Lassa fever. Sci Rep 8:5939

Boisen ML, Schieffelin JS, Goba A, Oottamasathien D, Jones AB, Shaffer JG, Hastie KM, Hartnett JN, Momoh M, Fullah M, Gabiki M, Safa S, Zandonatti M, Fusco M, Bornholdt Z, Abelson D, Gire SK, Andersen KG, Tariyal R, Stremlau M, Cross RW, Geisbert JB, Pitts KR, Geisbert TW, Kulakoski P, Wilson RB, Henderson L, Sabeti PC, Grant DS, Garry RF, Saphire EO, Branco LM, Khan SH (2015) Multiple circulating infections can mimic the early stages of viral hemorrhagic fevers and possible human exposure to filoviruses in Sierra Leone prior to the 2014 outbreak. Viral Immunol 28:19–31

Boisen ML, Uyigue E, Aiyepada J, Siddle KJ, Oestereich L, Nelson DKS, Bush DJ, Rowland MM, Heinrich ML, Eromon P, Kayode AT, Odia I, Adomeh DI, Muoebonam EB, Akhilomen P, Okonofua G, Osiemi B, Omoregie O, Airende M, Agbukor J, Ehikhametalor S, Aire CO, Duraffour S, Pahlmann M, Böhm W, Barnes KG, Mehta S, Momoh M, Sandi JD, Goba A, Folarin OA, Ogbaini-Emovan E, Asogun DA, Tobin EA, Akpede GO, Okogbenin SA, Okokhere PO, Grant DS, Schieffelin JS, Sabeti PC, Günther S, Happi CT, Branco LM, Garry RF (2020) Field evaluation of a Pan-Lassa rapid diagnostic test during the 2018 Nigerian Lassa fever outbreak. Sci Rep 10:8724

Bowen MD, Rollin PE, Ksiazek TG, Hustad HL, Bausch DG, Demby AH, Bajani MD, Peters CJ, Nichol ST (2000) Genetic diversity among Lassa virus strains. J Virol 74:6992–7004

Branco LM, Boisen ML, Andersen KG, Grove JN, Moses LM, Muncy IJ, Henderson LA, Schieffellin JS, Robinson JE, Bangura JJ, Grant DS, Raabe VN, Fonnie M, Zaitsev EM, Sabeti PC, Garry RF (2011a) Lassa hemorrhagic fever in a late term pregnancy from northern Sierra Leone with a positive maternal outcome: case report. Virol J 8:404

Branco LM, Grove JN, Boisen ML, Shaffer JG, Goba A, Fullah M, Momoh M, Grant DS, Garry RF (2011b) Emerging trends in Lassa fever: redefining the role of immunoglobulin M and inflammation in diagnosing acute infection. Virol J 8:478

Buba MI, Dalhat MM, Nguku PM, Waziri N, Mohammad JO, Bomoi IM, Onyiah AP, Onwujei J, Balogun MS, Bashorun AT, Nsubuga P, Nasidi A (2018) Mortality among confirmed Lassa fever cases during the 2015–2016 outbreak in Nigeria. Am J Public Health 108:262–264

Cashman KA, Wilkinson ER, Zeng X, Cardile AP, Facemire PR, Bell TM, Bearss JJ, Shaia CI, Schmaljohn CS (2018) Immune-mediated systemic vasculitis as the proposed cause of sudden-onset sensorineural hearing loss following lassa virus exposure in cynomolgus macaques. mBio 9

Cheng HY, French CE, Salam AP, Dawson S, McAleenan A, McGuinness LA, Savović J, Horby PW, Sterne JAC (2022) Lack of evidence for ribavirin treatment of Lassa fever in systematic review of published and unpublished studies(1). Emerg Infect Dis 28:1559–1568

Chiosi JJ, Schieffelin JS, Shaffer JG, Grant DS (2022) Evaluation of three clinical prediction tools to predict mortality in hospitalized patients with Lassa fever. Am J Trop Med Hyg 107:856–862

Cross RW, Hastie KM, Mire CE, Robinson JE, Geisbert TW, Branco LM, Ollmann Saphire E, Garry RF (2019) Antibody therapy for Lassa fever. Curr Opin Virol 37:97–104

Cross RW, Mire CE, Branco LM, Geisbert JB, Rowland MM, Heinrich ML, Goba A, Momoh M, Grant DS, Fullah M, Khan SH, Robinson JE, Geisbert TW, Garry RF (2016) Treatment of Lassa virus infection in outbred guinea pigs with first-in-class human monoclonal antibodies. Antiviral Res 133:218–222

Cummins D, Bennett D, Fisher-Hoch SP, Farrar B, McCormick JB (1989a) Electrocardiographic abnormalities in patients with Lassa fever. J Trop Med Hyg 92:350–355

Cummins D, Fisher-Hoch SP, Walshe KJ, Mackie IJ, McCormick JB, Bennett D, Perez G, Farrar B, Machin SJ (1989b) A plasma inhibitor of platelet aggregation in patients with Lassa fever. Br J Haematol 72:543–548

Dahmane A, van Griensven J, Van Herp M, Van den Bergh R, Nzomukunda Y, Prior J, Alders P, Jambai A, Zachariah R (2014) Constraints in the diagnosis and treatment of Lassa fever and the effect on mortality in hospitalized children and women with obstetric conditions in a rural district hospital in Sierra Leone. Trans R Soc Trop Med Hyg 108:126–132

Dalhat MM, Olayinka A, Meremikwu MM, Dan-Nwafor C, Iniobong A, Ntoimo LF, Onoh I, Mba S, Ohonsi C, Arinze C, Esu EB, Nwafor O, Oladipupo I, Onoja M, Ilori E, Okonofua F, Ochu CL, Igumbor EU, Adetifa I (2022) Epidemiological trends of Lassa fever in Nigeria, 2018–2021. PLoS ONE 17:e0279467

Downs IL, Shaia CI, Zeng X, Johnson JC, Hensley L, Saunders DL, Rossi F, Cashman KA, Esham HL, Gregory MK, Pratt WD, Trefry JC, Everson KA, Larcom CB, Okwesili AC, Cardile AP, Honko A (2020) Natural history of aerosol induced lassa fever in non human primates. Viruses 12

Duvignaud A, Jaspard M, Etafo IC, Serra B, Abejegah C, Gabillard D, Doutchi M, Alabi JF, Adedokun MA, Akinpelu AO, Oyegunle OO, Etafo J, Dede AO, Onyechi MN, Ireneh MU, Gbenga-Ayeni O, Fadiminiyi KG, Ehigbor PI, Ouattara E, Levy-Marchal C, Kärcher S, N'Guessan-Koffi L, Ahyi I, Amani E, Diabate M, Siloue B, Schaeffer J, Augier A, Ogbaini-Emovon E, Salam AP, Horby P, Ahmed LA, Gunther S, Adedosu AN, Anglaret X, Ayodeji OO, Malvy D (2020) Lassa fever clinical course and setting a standard of care for future randomized trials: a protocol for a cohort study of Lassa-infected patients in Nigeria (LASCOPE). Travel Med Infect Dis:101557

Eberhardt KA, Mischlinger J, Jordan S, Groger M, Günther S, Ramharter M (2019) Ribavirin for the treatment of Lassa fever: a systematic review and meta-analysis. Int J Infect Dis IJID Official Publ Int Soc Infect Dis 87:15–20

Edington GM, White HA (1972) The pathology of Lassa fever. Trans R Soc Trop Med Hyg 66:381–389

Ehichioya DU, Asogun DA, Ehimuan J, Okokhere PO, Pahlmann M, Olschlager S, Becker-Ziaja B, Gunther S, Omilabu SA (2012) Hospital-based surveillance for Lassa fever in Edo State, Nigeria, 2005–2008. Trop Med Int Health 17:1001–1004

Ferm VH, Willhite C, Kilham L (1978) Teratogenic effects of ribavirin on hamster and rat embryos. Teratology 17:93–101

Ficenec SC, Percak J, Arguello S, Bays A, Goba A, Gbakie M, Shaffer JG, Emmett SD, Schieffelin JS, Bausch D (2020) Lassa fever induced hearing loss: the neglected disability of hemorrhagic fever. Int J Infect Dis IJID Official Publ Int Soc Infect Dis 100:82–87

Ficenec SC, Schieffelin JS, Emmett SD (2019) A review of hearing loss associated with Zika, Ebola, and Lassa fever. Am J Trop Med Hyg 101:484–490

Fichet-Calvet E, Rogers DJ (2009) Risk maps of Lassa fever in West Africa. PLoS Negl Trop Dis 3:e388

Fisher-Hoch S, McCormick JB, Sasso D, Craven RB (1988) Hematologic dysfunction in Lassa fever. J Med Virol 26:127–135

Fisher-Hoch SP, Gborie S, Parker L, Huggins J (1992) Unexpected adverse reactions during a clinical trial in rural West Africa. Antiviral Res 19:139–147

Fisher-Hoch SP, McCormick JB (1987) Pathophysiology and treatment of Lassa fever. Curr Top Microbiol Immunol 134:231–239

Fisher-Hoch SP, Mitchell SW, Sasso DR, Lange JV, Ramsey R, McCormick JB (1987) Physiological and immunologic disturbances associated with shock in a primate model of Lassa fever. J Infect Dis 155:465–474

Fisher-Hoch SP, Price ME, Craven RB, Price FM, Forthall DN, Sasso DR, Scott SM, McCormick JB (1985) Safe intensive-care management of a severe case of Lassa fever with simple barrier nursing techniques. Lancet 2:1227–1229

Fisher-Hoch SP, Tomori O, Nasidi A, Perez-Oronoz GI, Fakile Y, Hutwagner L, McCormick JB (1995) Review of cases of nosocomial Lassa fever in Nigeria: the high price of poor medical practice. BMJ 311:857–859

Flatz L, Rieger T, Merkler D, Bergthaler A, Regen T, Schedensack M, Bestmann L, Verschoor A, Kreutzfeldt M, Bruck W, Hanisch UK, Gunther S, Pinschewer DD (2010) T cell-dependence of Lassa fever pathogenesis. PLoS Pathog 6:e1000836

Frame JD (1989) Clinical features of Lassa fever in Liberia. Rev Infect Dis 11(Suppl 4):S783-789

Frame JD, Verbrugge GP, Gill RG, Pinneo L (1984) The use of Lassa fever convalescent plasma in Nigeria. Trans R Soc Trop Med Hyg 78:319–324

Gale TV, Horton TM, Grant DS, Garry RF (2017) Metabolomics analyses identify platelet activating factors and heme breakdown products as Lassa fever biomarkers. PLoS Negl Trop Dis 11:e0005943

Gale TV, Schieffelin JS, Branco LM, Garry RF, Grant DS (2020) Elevated L-threonine is a biomarker for Lassa fever and Ebola. Virol J 17:188

Garry RF (2023) Lassa fever—the road ahead. Nat Rev Microbiol 21:87–96

Gary JM, Welch SR, Ritter JM, Coleman-McCray J, Huynh T, Kainulainen MH, Bollweg BC, Parihar V, Nichol ST, Zaki SR, Spiropoulou CF, Spengler JR (2019) Lassa virus targeting of anterior uvea and endothelium of cornea and conjunctiva in eye of guinea pig model. Emerg Infect Dis 25:865–874

Goba A, Khan SH, Fonnie M, Fullah M, Moigboi A, Kovoma A, Sinnah V, Yoko N, Rogers H, Safai S, Momoh M, Koroma V, Kamara FK, Konowu E, Yillah M, French I, Mustapha I, Kanneh F, Foday M, McCarthy H, Kallon T, Kallon M, Naiebu J, Sellu J, Jalloh AA, Gbakie M, Kanneh L, Massaly JL, Kargbo D, Kargbo B, Vandi M, Gbetuwa M, Gevao SM, Sandi JD, Jalloh SC, Grant DS, Blyden SO, Crozier I, Schieffelin JS, McLellan SL, Jacob, ST, Boisen ML, Hartnett JN, Cross RW, Branco LM, Andersen KG, Yozwiak NL, Gire SK, Tariyal R, Park DJ, Haislip AM, Bishop CM, Melnik LI, Gallaher WR, Wimley WC, He J, Shaffer JG, Sullivan BM, Grillo S, Oman S, Garry CE, Edwards DR, McCormick SJ, Elliott DH, Rouelle JA, Kannadka CB, Reyna AA, Bradley BT, Yu H, Yenni RE, Hastie KM, Geisbert JB, Kulakosky PC, Wilson RB, Oldstone MB, Pitts KR, Henderson LA, Robinson JE, Geisbert TW, Saphire EO, Happi CT, Asogun DA, Sabeti PC, Garry RF (2016) An outbreak of ebola virus disease in the Lassa fever zone. J Infect Dis

Grant D, Samuels R, Engel E, Alhasan F, Gbakie M, Goba A, Kanneh L, Koroma V, Momoh M, Sandi J, Bond N, Levy D, McLaughlin A, Nelson M, Yerkes N, Boisen M, Branco L, Varley C, Bausch D, Shaffer J, Garry R, Schieffelin J (2023a) Stage of disease, not time to presentation, predicts outcome among Lassa fever patients in Eastern Sierra Leone, submitted

Grant DS, Engel EJ, Roberts Yerkes N, Kanneh L, Koninga J, Gbakie MA, Alhasan F, Kanneh FB, Kanneh IM, Kamara FK, Momoh M, Yillah MS, Foday M, Okoli A, Zeoli A, Weldon C, Bishop CM, Zheng C, Hartnett J, Chao K, Shore K, Melnik LI, Mucci M, Bond NG, Doyle P, Yenni R, Podgorski R, Ficenec SC, Moses L, Shaffer JG, Garry RF, Schieffelin JS (2023b) Seroprevalence of anti-Lassa virus IgG antibodies in three districts of Sierra Leone: a cross-sectional, population-based study. PLoS Negl Trop Dis 17:e0010938

Groger M, Akhideno P, Kleist CJ, Babatunde FO, Edeawe O, Hinzmann J, Akhigbe T, Nwatuzor J, Eifediyi G, Müller J, Hinrichs M, Pahlmann M, Sarpong FN, Wagner C, Thielebein A, Aihonwalan L, Koch T, Riedner M, Ogbaini-Emovon E, Okogbenin S, Günther S, Wicha SG, Ramharter M, Oestereich L, Duraffour S, Erameh C (2023) Pharmacokinetics of ribavirin in the treatment of Lassa fever: an observational clinical study at the Irrua specialist teaching hospital, Edo State, Nigeria. Clin Infect Dis 76:e841–e848

Grove JN, Branco LM, Boisen ML, Muncy IJ, Henderson LA, Schieffellin JS, Robinson JE, Bangura JJ, Fonnie M, Schoepp RJ, Hensley LE, Seisay A, Fair JN, Garry RF (2011) Capacity building permitting comprehensive monitoring of a severe case of Lassa hemorrhagic fever in Sierra Leone with a positive outcome: case report. Virol J 8:314

Hadi CM, Goba A, Khan SH, Bangura J, Sankoh M, Koroma S, Juana B, Bah A, Coulibaly M, Bausch DG (2010) Ribavirin for Lassa fever postexposure prophylaxis. Emerg Infect Dis 16:2009–2011

Hansen F, Jarvis MA, Feldmann H, Rosenke K (2021) Lassa virus treatment options. Microorganisms 9

Happi AN, Ogunsanya OA, Oguzie JU, Oluniyi PE, Olono AS, Heeney JL, Happi CT (2021) Microbial metagenomic approach uncovers the first rabbit haemorrhagic disease virus genome in Sub-Saharan Africa. Sci Rep 11:13689

Hartnett JN, Boisen ML, Oottamasathien D, Jones AB, Millett MM, Nelson DS, Muncy IJ, Goba A, Momoh M, Fullah M, Mire CE, Geisbert JB, Geisbert TW, Holton DL, Rouelle JA, Kannadka CB, Reyna AA, Moses LM, Khan SH, Gevao SM, Grant DS, Robinson JE, Happi C, Pitts KR, Garry RF, Branco LM, Consortium TVHF (2015) Current and emerging strategies for the diagnosis, prevention and treatment of Lassa fever. Future Virol 10:559–584

Hastie KM, Kimberlin CR, Zandonatti MA, MacRae IJ, Saphire EO (2011) Structure of the Lassa virus nucleoprotein reveals a dsRNA-specific 3′ to 5′ exonuclease activity essential for immune suppression. Proc Natl Acad Sci USA 108:2396–2401

Hayes MW, Carrion R Jr, Nunneley J, Medvedev AE, Salvato MS, Lukashevich IS (2012) Pathogenic old world arenaviruses inhibit TLR2/Mal-dependent proinflammatory cytokines in vitro. J Virol 86:7216–7226

Heinrich ML, Boisen ML, Nelson DKS, Bush DJ, Cross RW, Koval AP, Hoffmann AR, Beddingfield BJ, Hastie KM, Rowland MM, Aimukanova I, Koval S, Lathigra R, Borisevich V, Momoh M, Sandi JD, Goba A, Odia L, Baimba F, Aiyepada JO, Ebo B, Eromon P, Ugwu C, Folarin O, Olumade T, Onyechi MN, Etafo J, Adeyemi R, Ella EE, Aminu M, Gomerep SS, Eke MA, Ogunsanya O, Akpede GO, Asogun DO, Okogbenin SA, Okokhere PO, Holst J, Shaffer JG, Schieffelin JS, Geisbert TW, Saphire EO, Happi CT, Grant DS, Garry RF, Branco LM (2020) Antibodies from Sierra Leonean and Nigerian Lassa fever survivors cross-react with recombinant proteins representing Lassa viruses of divergent lineages. Sci Rep 10:16030

Hensley LE, Smith MA, Geisbert JB, Fritz EA, Daddario-DiCaprio KM, Larsen T, Geisbert TW (2011) Pathogenesis of Lassa fever in cynomolgus macaques. Virol J 8:205

Horton LE, Cross RW, Hartnett JN, Engel EJ, Sakabe S, Goba A, Momoh M, Sandi JD, Geisbert TW, Garry RF, Schieffelin JS, Grant DS, Sullivan BM (2020) Endotheliopathy and platelet dysfunction as hallmarks of fatal Lassa fever. Emerg Infect Dis 26:2625–2637

Huynh T, Gary JM, Welch SR, Coleman-McCray J, Harmon JR, Kainulainen MH, Bollweg BC, Ritter JM, Shieh WJ, Nichol ST, Zaki SR, Spiropoulou CF, Spengler JR (2020) Lassa virus antigen distribution and inflammation in the ear of infected strain 13/N Guinea pigs. Antiviral Res 183:104928

Ibekwe TS, Nwegbu MM, Asogun D, Adomeh DI, Okokhere PO (2012) The sensitivity and specificity of Lassa virus IgM by ELISA as screening tool at early phase of Lassa fever infection. Niger Med J 53:196–199

Ibekwe TS, Okokhere PO, Asogun D, Blackie FF, Nwegbu MM, Wahab KW, Omilabu SA, Akpede GO (2011) Early-onset sensorineural hearing loss in Lassa fever. Euro Arch Oto-Rhino-Laryngol Official J Euro Federation Oto-Rhino-Laryngol Soc (EUFOS) Affil German Soc Oto-Rhino-Laryngol Head Neck Surgery 268:197–201

Ijarotimi IT, Ilesanmi OS, Aderinwale A, Abiodun-Adewusi O, Okon IM (2018) Knowledge of Lassa fever and use of infection prevention and control facilities among health care workers during Lassa fever outbreak in Ondo State, Nigeria. Pan Afr Med J 30:56

Ilori EA, Furuse Y, Ipadeola OB, Dan-Nwafor CC, Abubakar A, Womi-Eteng OE, Ogbaini-Emovon E, Okogbenin S, Unigwe U, Ogah E, Ayodeji O, Abejegah C, Liasu AA, Musa EO, Woldetsadik SF, Lasuba CLP, Alemu W, Ihekweazu C (2019) Epidemiologic and Clinical Features of Lassa fever outbreak in Nigeria, January 1-May 6, 2018. Emerg Infect Dis 25:1066–1074

Jahrling PB, Frame JD, Rhoderick JB, Monson MH (1985) Endemic Lassa fever in Liberia. IV. Selection of optimally effective plasma for treatment by passive immunization. Trans R Soc Trop Med Hyg 79:380–384

Jahrling PB, Peters CJ (1984) Passive antibody therapy of Lassa fever in cynomolgus monkeys: importance of neutralizing antibody and Lassa virus strain. Infect Immun 44:528–533

Jahrling PB, Peters CJ, Stephen EL (1984) Enhanced treatment of Lassa fever by immune plasma combined with ribavirin in cynomolgus monkeys. J Infect Dis 149:420–427

Jetoh RW, Malik S, Shobayo B, Taweh F, Yeabah TO, George J, Gbelee B, Teahton J, Jaryan F, Tegli M, Umeokonkwo CD, MaCauley J (2022) Epidemiological characteristics of Lassa fever cases in Liberia: a retrospective analysis of surveillance data, 2019–2020. Int J Infect Dis IJID Official Publ Int Soc Infect Dis 122:767–774

Kayem ND, Benson C, Aye CYL, Barker S, Tome M, Kennedy S, Ariana P, Horby P (2020) Lassa fever in pregnancy: a systematic review and meta-analysis. Trans R Soc Trop Med Hyg 114:385–396

Kenmoe S, Tchatchouang S, Ebogo-Belobo JT, Ka'e AC, Mahamat G, Guiamdjo Simo RE, Bowo-Ngandji A, Demeni Emoh CP, Che E, Tchami Ngongang D, Amougou-Atsama M, Nzukui ND, Mbongue Mikangue CA, Mbaga DS, Kenfack S, Kingue Bebey SR, Amvongo Adjia N, Efietngab AN, Tazokong HR, Modiyinji AF, Kengne-Nde C, Sadeuh-Mba SA, Njouom R (2020) Systematic review and meta-analysis of the epidemiology of Lassa virus in humans, rodents and other mammals in sub-Saharan Africa. PLoS Negl Trop Dis 14:e0008589

Khan SH, Goba A, Chu M, Roth C, Healing T, Marx A, Fair J, Guttieri MC, Ferro P, Imes T, Monagin C, Garry RF, Bausch DG (2008) New opportunities for field research on the pathogenesis and treatment of Lassa fever. Antiviral Res 78:103–115

Klitting R, Kafetzopoulou LE, Thiery W, Dudas G, Gryseels S, Kotamarthi A, Vrancken B, Gangavarapu K, Momoh M, Sandi JD, Goba A, Alhasan F, Grant DS, Okogbenin S, Ogbaini-Emovo E, Garry RF, Smither AR, Zeller M, Pauthner MG, McGraw M, Hughes LD, Duraffour S, Günther S, Suchard MA, Lemey P, Andersen KG, Dellicour S (2022) Predicting the evolution of the Lassa virus endemic area and population at risk over the next decades. Nat Commun 13:5596

Kofman A, Choi MJ, Rollin PE (2019) Lassa fever in travelers from West Africa, 1969–2016. Emerg Infect Dis 25:245–248

LaVergne SM, Sakabe S, Momoh M, Kanneh L, Bond N, Garry RF, Grant DS, de la Torre JC, Oldstone MBA, Schieffelin JS, Sullivan BM (2022) Expansion of CD8+ T cell population in Lassa virus survivors with low T cell precursor frequency reveals durable immune response in most survivors. PLoS Negl Trop Dis 16:e0010882

Lecompte E, Fichet-Calvet E, Daffis S, Koulemou K, Sylla O, Kourouma F, Dore A, Soropogui B, Aniskin V, Allali B, Kouassi Kan S, Lalis A, Koivogui L, Gunther S, Denys C, ter Meulen J (2006) Mastomys natalensis and Lassa fever, West Africa. Emerg Infect Dis 12:1971–1974

Leifer E, Gocke DJ, Bourne H (1970) Lassa fever, a new virus disease of man from West Africa. II. Report of a laboratory-acquired infection treated with plasma from a person recently recovered from the disease. Am J Trop Med Hyg 19:677–679

Li AL, Grant D, Gbakie M, Kanneh L, Mustafa I, Bond N, Engel E, Schieffelin J, Vandy MJ, Yeh S, Shantha JG (2020) Ophthalmic manifestations and vision impairment in Lassa fever survivors. PLoS ONE 15:e0243766

Li H, Buck T, Zandonatti M, Yin J, Moon-Walker A, Fang J, Koval A, Heinrich ML, Rowland MM, Diaz Avalos R, Schendel SL, Parekh D, Zyla D, Enriquez A, Harkins S, Sullivan B, Smith V, Chukwudozie O, Watanabe R, Robinson JE, Garry RF, Branco LM, Hastie KM, Saphire EO (2022) A cocktail of protective antibodies subverts the dense glycan shield of Lassa virus. Sci Transl Med 14:eabq0991

Lukashevich IS, Paessler S, de la Torre JC (2019) Lassa virus diversity and feasibility for universal prophylactic vaccine. F1000Research 8

Mateer EJ, Huang C, Shehu NY, Paessler S (2018) Lassa fever-induced sensorineural hearing loss: a neglected public health and social burden. PLoS Negl Trop Dis 12:e0006187

McCormick JB, Fisher-Hoch SP (2002) Lassa fever. Curr Top Microbiol Immunol 262:75–109

McCormick JB, King IJ, Webb PA, Johnson KM, O'Sullivan R, Smith ES, Trippel S, Tong TC (1987) A case-control study of the clinical diagnosis and course of Lassa fever. J Infect Dis 155:445–455

McCormick JB, King IJ, Webb PA, Scribner CL, Craven RB, Johnson KM, Elliott LH, Belmont-Williams R (1986) Lassa fever. Effective therapy with ribavirin. N Engl J Med 314:20–26

Melnik LI (2022) Lassa virus countermeasures. Curr Top Microbiol Immunol

Mendenhall M, Russell A, Smee DF, Hall JO, Skirpstunas R, Furuta Y, Gowen BB (2011) Effective oral favipiravir (T-705) therapy initiated after the onset of clinical disease in a model of arenavirus hemorrhagic fever. PLoS Negl Trop Dis 5:e1342

Mertens PE, Patton R, Baum JJ, Monath TP (1973) Clinical presentation of Lassa fever cases during the hospital epidemic at Zorzor, Liberia, March-April 1972. Am J Trop Med Hyg 22:780–784

Meyer B, Ly H (2016) Inhibition of innate immune responses is key to pathogenesis by arenaviruses. J Virol 90:3810–3818

Mire CE, Cross RW, Geisbert JB, Borisevich V, Agans KN, Deer DJ, Heinrich ML, Rowland MM, Goba A, Momoh M, Boisen ML, Grant DS, Fullah M, Khan SH, Fenton KA, Robinson JE, Branco LM, Garry RF, Geisbert TW (2017) Human-monoclonal-antibody therapy protects nonhuman primates against advanced Lassa fever. Nat Med 23:1146–1149

Monath TP (2019) A short history of Lassa fever: the first 10–15 years after discovery. Curr Opin Virol 37:77–83

Monath TP, Maher M, Casals J, Kissling RE, Cacciapuoti A (1974) Lassa fever in the Eastern Province of Sierra Leone, 1970–1972. II. Clinical observations and virological studies on selected hospital cases. Am J Trop Med Hyg 23:1140–1149

Monson MH, Cole AK, Frame JD, Serwint JR, Alexander S, Jahrling PB (1987) Pediatric Lassa fever: a review of 33 Liberian cases. Am J Trop Med Hyg 36:408–415

Nagata T, Lefor AK, Hasegawa M, Ishii M (2015) Favipiravir: a new medication for the Ebola virus disease pandemic. Disaster Med Publ Health Prep 9:79–81

Njuguna C, Vandi M, Liyosi E, Githuku J, Squire JS, Njeru I, Rufus I, Katawera V, Gachari W, Musoke R, Amuzu C, Fahnbulleh M, Bunting-Graden J, Kayita J, Bunn J, Talisuna A, Yoti Z (2022a) After action review of the response to an outbreak of Lassa fever in Sierra Leone, 2019: Best practices and lessons learnt. PLoS Negl Trop Dis 16:e0010755

Njuguna C, Vandi M, Liyosi E, Githuku J, Wurie A, Njeru I, Raftery P, Amuzu C, Maruta A, Musoke R, Fahnbulleh M, Bunting J, Gbandeh S, Talisuna A, Yoti Z (2022b) A challenging response to a Lassa fever outbreak in a non endemic area of Sierra Leone in 2019 with export of cases to The Netherlands. Int J Infect Dis IJID Official Publ Int Soc Infect Dis 117:295–301

Oguzie JU, Nwangwu UC, Oluniyi PE, Olumade TJ, George UE, Kazeem A, Bankole BE, Brimmo FO, Asadu CC, Chukwuekezie OC, Ochu JC, Makwe CO, Dogunro FA, Onwude CO, Nwachukwu WE, Ezihe EK, Okonkwo GK, Umazi NE, Maikere J, Agashi NO, Eloy EI, Anokwu SO, Okoronkwo AI, Nwosu EM, Etiki SO, Ngwu IM, Ihekweazu C, Folarin OA, Komolafe IOO, Happi CT (2022) Metagenomic sequencing characterizes a wide diversity of viruses in field mosquito samples in Nigeria. Sci Rep 12:7616

Okogbenin S, Okoeguale J, Akpede G, Colubri A, Barnes KG, Mehta S, Eifediyi R, Okogbo F, Eigbefoh J, Momoh M, Rafiu M, Adomeh D, Odia I, Aire C, Atafo R, Okonofua M, Pahlman M, Becker-Ziaja B, Asogun D, Okokhere P, Happi C, Gunther S, Sabeti PC, Ogbaini-Emovon E (2019) Retrospective cohort study of Lassa fever in pregnancy, Southern Nigeria. Emerg Infect Dis 25

Okokhere P, Colubri A, Azubike C, Iruolagbe C, Osazuwa O, Tabrizi S, Chin E, Asad S, Ediale E, Rafiu M, Adomeh D, Odia I, Atafo R, Aire C, Okogbenin S, Pahlman M, Becker-Ziaja B, Asogun D, Fradet T, Fry B, Schaffner SF, Happi C, Akpede G, Gunther S, Sabeti PC (2018a) Clinical and laboratory predictors of Lassa fever outcome in a dedicated treatment facility in Nigeria: a retrospective, observational cohort study. Lancet Infect Dis 18:684–695

Okokhere PO, Erameh CO, Alikah F, Akhideno PE, Iruolagbe CO, Osazuwa OO, Bankole IA, Adomeh DI, Odia I, Aire CO, Oyakhilome J, Agbukor J, Atafo RO, Asogun DA, Happi CT, Sabeti PC, Gunther S, Akpede GO (2018b) Acute Lassa virus encephalitis with lassa virus in the cerebrospinal fluid but absent in the blood: a case report with a positive outcome. Case Rep Neurol 10:150–158

Okokhere PO, Ibekwe TS, Akpede GO (2009) Sensorineural hearing loss in Lassa fever: two case reports. J Med Case Reports 3:36

Olatunde LO, Owhin SO, Momoh AJ, Adebayo TT, Babatunde E, Amodu BE, Ogunmodede AF, Kareem AO, Amenkhienan IF, Abidoye AT, Ashaju-Kayode OC, Jaiyeola OO, Ezeigbo PO, Alabi JF, Ahmed LA (2021) Predictors of in-hospital mortality in patients with RT-PCR confirmed Lassa fever infection treated at a national treatment center, South West Nigeria. Clin Immunol Res 5:1–5

Oldstone MB (2002) Arenaviruses. I. The epidemiology molecular and cell biology of arenaviruses. Introduction. Curr Top Microbiol Immunol 262:V–xii

Olschlager S, Gunther S (2012) Rapid and specific detection of Lassa virus by reverse transcription-PCR coupled with oligonucleotide array hybridization. J Clin Microbiol 50:2496–2499

Port JR, Wozniak DM, Oestereich L, Pallasch E, Becker-Ziaja B, Müller J, Rottstegge M, Olal C, Gómez-Medina S, Oyakhliome J, Ighodalo Y, Omomoh E, Olokor T, Adomeh DI, Asogun D, Ogbani-Emovon E, Hartmann K, Krasemann S, Nelson EV, Escudero-Pérez B, McElroy AK, Günther S, Muñoz-Fontela C (2020) Severe human Lassa fever is characterized by nonspecific T-cell activation and lymphocyte homing to inflamed tissues. J Virol 94

Price ME, Fisher-Hoch SP, Craven RB, McCormick JB (1988) A prospective study of maternal and fetal outcome in acute Lassa fever infection during pregnancy. BMJ 297:584–587

Raabe VN, Kann G, Ribner BS, Morales A, Varkey JB, Mehta AK, Lyon GM, Vanairsdale S, Faber K, Becker S, Eickmann M, Strecker T, Brown S, Patel K, De Leuw P, Schuettfort G, Stephan C, Rabenau H, Klena JD, Rollin PE, McElroy A, Stroher U, Nichol S, Kraft CS, Wolf T (2017) Favipiravir and ribavirin treatment of epidemiologically linked cases of Lassa fever. Clin Infect Dis 65:855–859

Rigby I, Michelen M, Dagens A, Cheng V, Dahmash D, Harriss E, Webb E, Cai E, Lipworth S, Oti A, Balan V, Piotrowski H, Nartowski R, Rojek A, Groves H, Hart P, Cevik M, Bosa HK, Blumberg L, Fletcher TE, Horby PW, Jacob ST, Sigfrid L (2023) Standard of care for viral haemorrhagic fevers (VHFs): a systematic review of clinical management guidelines for high-priority VHFs. Lancet Infect Dis

Robinson JE, Hastie KM, Cross RW, Yenni RE, Elliott DH, Rouelle JA, Kannadka CB, Smira AA, Garry CE, Bradley BT, Yu H, Shaffer JG, Boisen ML, Hartnett JN, Zandonatti MA, Rowland MM, Heinrich ML, Martinez-Sobrido L, Cheng B, de la Torre JC, Andersen KG, Goba A, Momoh M, Fullah M, Gbakie M, Kanneh L, Koroma VJ, Fonnie R, Jalloh SC, Kargbo B, Vandi MA, Gbetuwa M, Ikponmwosa O, Asogun DA, Okokhere PO, Follarin OA, Schieffelin JS, Pitts KR, Geisbert JB, Kulakoski PC, Wilson RB, Happi CT, Sabeti PC, Gevao SM, Khan SH, Grant DS, Geisbert TW, Saphire EO, Branco LM, Garry RF (2016) Most neutralizing human monoclonal antibodies target novel epitopes requiring both Lassa virus glycoprotein subunits. Nat Commun 7:11544

Roth PJ, Grant DS, Ngegbai AS, Schieffelin J, McClelland RS, Jarrett OD (2015) Factors associated with mortality in febrile patients in a government referral hospital in the Kenema district of Sierra Leone. Am J Trop Med Hyg 92:172–177

Sabeti PC, Schaffner SF, Fry B, Lohmueller J, Varilly P, Shamovsky O, Palma A, Mikkelsen TS, Altshuler D, Lander ES (2006) Positive natural selection in the human lineage. Science 312:1614–1620

Safronetz D, Rosenke K, Westover JB, Martellaro C, Okumura A, Furuta Y, Geisbert J, Saturday G, Komeno T, Geisbert TW, Feldmann H, Gowen BB (2015) The broad-spectrum antiviral favipiravir protects guinea pigs from lethal Lassa virus infection post-disease onset. Sci Rep 5:14775

Safronetz D, Strong JE, Feldmann F, Haddock E, Sogoba N, Brining D, Geisbert TW, Scott DP, Feldmann H (2013) A recently isolated Lassa virus from Mali demonstrates atypical clinical disease manifestations and decreased virulence in cynomolgus macaques. J Infect Dis 207:1316–1327

Sakabe S, Hartnett JN, Ngo N, Goba A, Momoh M, Demby Sandi J, Kanneh L, Cubitt B, Garcia SD, Ware BC, Kotliar D, Robles-Sikisaka R, Gangavarapu K, Branco LM, Eromon P, Odia I, Ogbaini-Emovon E, Folarin O, Okogbenin S, Okokhere PO, Happi C, Sabeti PC, Andersen KG, Garry RF, de la Torre JC, Grant DS, Schieffelin JS, Oldstone MBA, Sullivan BM (2020) Identification of common CD8+ T cell epitopes from Lassa fever survivors in Nigeria and Sierra Leone. J Virol

Salam AP, Cheng V, Edwards T, Olliaro P, Sterne J, Horby P (2021) Time to reconsider the role of ribavirin in Lassa fever. PLoS Negl Trop Dis 15:e0009522

Salam AP, Duvignaud A, Jaspard M, Malvy D, Carroll M, Tarning J, Olliaro PL, Horby PW (2022) Ribavirin for treating Lassa fever: a systematic review of pre-clinical studies and implications for human dosing. PLoS Negl Trop Dis 16:e0010289

Saleh F, Popoola BO, Arinze C, Elisha AA, Dan-Nwafor C, Olajide L, Ilori E, Ochu CL, Ihekweazu CA (2022) Adapting public health response through lessons learnt: Nigeria's experience from Lassa fever and COVID-19. BMJ Global Health 7

Samuels RJ, Moon TD, Starnes JR, Alhasan F, Gbakie M, Goba A, Koroma V, Momoh M, Sandi JD, Garry RF, Engel EJ, Shaffer JG, Schieffelin JS, Grant DS (2020) Lassa fever among children in eastern province, Sierra Leone: A 7-year retrospective analysis (2012–2018). Am J Trop Med Hyg 104:585–592

Schieffelin J, Samuels R, Goba A, Alhasan F, Gbakie M, Kanneh L, Koroma V, Momoh M, Sandi J, Bond N, Chiosi J, Engel E, Levy D, McLaughlin A, Nelson M, Yerkes N, Boisen M, Branco L, Varley C, Bausch D, Shaffer J, Garry R, Grant D (2023) Clinical presentation and renal dysfunction of acute and sub-acute Lassa fever in Sierra Leone, submitted

Shaffer JG, Grant DS, Schieffelin JS, Boisen ML, Goba A, Hartnett JN, Levy DC, Yenni RE, Moses LM, Fullah M, Momoh M, Fonnie M, Fonnie R, Kanneh L, Koroma VJ, Kargbo K, Ottomassathien D, Muncy IJ, Jones AB, Illick MM, Kulakosky PC, Haislip AM, Bishop CM, Elliot DH, Brown BL, Zhu H, Hastie KM, Andersen KG, Gire SK, Tabrizi S, Tariyal R, Stremlau M, Matschiner A, Sampey DB, Spence JS, Cross RW, Geisbert JB, Folarin OA, Happi CT, Pitts KR, Geske FJ, Geisbert TW, Saphire EO, Robinson JE, Wilson RB, Sabeti PC, Henderson LA, Khan SH, Bausch DG, Branco LM, Garry RF (2014) Lassa fever in post-conflict Sierra Leone. PLoS Negl Trop Dis 8:e2748

Shaffer JG, Schieffelin JS, Grant DS, Goba A, Momoh M, Kanneh L, Levy DC, Hartnett JN, Boisen ML, Branco LM, Garry RF (2019) Data set on Lassa fever in post-conflict Sierra Leone. Data Brief 23:103673

Shaffer JG, Schieffelin JS, Momoh M, Goba A, Kanneh L, Alhasan F, Gbakie M, Engel EJ, Bond NG, Hartnett JN, Nelson DKS, Bush DJ, Boisen ML, Heinrich ML, Rowland MM, Branco LM, Samuels RJ, Garry RF, Grant DS, The Viral Hemorrhagic Fever, C (2021) Space-time trends in Lassa fever in Sierra Leone by ELISA Serostatus, 2012–2019. Microorganisms 9

Shao J, Liang Y, Ly H (2015) Human hemorrhagic fever causing arenaviruses: molecular mechanisms contributing to virus virulence and disease pathogenesis. Pathogens 4:283–306

Shieh W-J, Demby A, Jones T, Goldsmith CS, Rollin PE, Ksiazek TG, Peters CJ, Zaki SR (2021) Pathology and pathogenesis of Lassa fever: novel immunohistochemical findings in fatal cases and clinico-pathologic correlation. Clin Infect Dis 74:1821–1830

Siddle KJ, Eromon P, Barnes KG, Mehta S, Oguzie JU, Odia I, Schaffner SF, Winnicki SM, Shah RR, Qu J, Wohl S, Brehio P, Iruolagbe C, Aiyepada J, Uyigue E, Akhilomen P, Okonofua G, Ye S, Kayode T, Ajogbasile F, Uwanibe J, Gaye A, Momoh M, Chak B, Kotliar D, Carter A, Gladden-Young A, Freije CA, Omoregie O, Osiemi B, Muoebonam EB, Airende M, Enigbe R, Ebo B, Nosamiefan I, Oluniyi P, Nekoui M, Ogbaini-Emovon E, Garry RF, Andersen KG, Park DJ, Yozwiak NL, Akpede G, Ihekweazu C, Tomori O, Okogbenin S, Folarin OA, Okokhere PO, MacInnis BL, Sabeti PC, Happi CT (2018) Genomic analysis of Lassa virus during an increase in cases in Nigeria in 2018. N Engl J Med 379:1745–1753

Simmons FB (1993) Lassa fever and sudden hearing loss. Otolaryngol Head Neck Surg 109:559

Smither AR, Bell-Kareem AR (2021) Ecology of Lassa virus. Curr Top Microbiol Immunol

Sogoba N, Feldmann H, Safronetz D (2012) Lassa fever in West Africa: evidence for an expanded region of endemicity. Zoonoses Publ Health 59(Suppl 2):43–47

Stein DR, Warner BM, Audet J, Soule G, Siragam V, Sroga P, Griffin BD, Leung A, Grolla A, Tierney K, Albietz A, Kobasa D, Musa AS, Ahmad A, Akinpelu AM, Mba N, Rosenke R, Scott DP, Saturday G, Ihekweazu C, Safronetz D (2021) Differential pathogenesis of closely related 2018 Nigerian outbreak clade III Lassa virus isolates. PLoS Pathog 17:e1009966

Sullivan BM, Sakabe S, Hartnett JN, Ngo N, Goba A, Momoh M, Demby Sandi J, Kanneh L, Cubitt B, Garcia SD, Ware BC, Kotliar D, Robles-Sikisaka R, Gangavarapu K, Branco L, Eromon P, Odia I, Ogbaini-Emovon E, Folarin O, Okogbenin S, Okokhere PO, Happi C, de la Torre JC, Sabeti PC, Andersen KG, Garry RF, Grant DS, Schieffelin JS, Oldstone MBA (2020) High crossreactivity of human T cell responses between Lassa virus lineages. PLoS Pathog 16: e1008352

Tessema SK, Inzaule SC, Christoffels A, Kebede Y, de Oliveira T, Ouma AEO, Happi CT, Nkengasong JN (2020) Accelerating genomics-based surveillance for COVID-19 response in Africa. Lancet Microbe 1:e227–e228

Usifoh SF, Odigie AE, Ighedosa SU, Uwagie-Ero EA, Aighewi IT (2019) Lassa fever-associated stigmatization among staff and students of the University of Benin, Nigeria. J Epidemiol Glob Health 9:107–115

Walker DH, McCormick JB, Johnson KM, Webb PA, Komba-Kono G, Elliott LH, Gardner JJ (1982) Pathologic and virologic study of fatal Lassa fever in man. Am J Pathol 107:349–356

Walls B (1985) Lassa fever and pregnancy. Midwives Chron 1168:136–138

Webb PA, McCormick JB, King IJ, Bosman I, Johnson KM, Elliott LH, Kono GK, O'Sullivan R (1986) Lassa fever in children in Sierra Leone, West Africa. Trans R Soc Trop Med Hyg 80:577–582

WHO (2016) Clinical management of patients with viral haemorrhagic fever. A pocket guide for front-line health workers. WHO Library, Geneva

WHO, 2022. Lassa fever

Winn WC Jr, Walker DH (1975) The pathology of human Lassa fever. Bull World Health Organ 52:535–545

Wolf T, Ellwanger R, Goetsch U, Wetzstein N, Gottschalk R (2020) Fifty years of imported Lassa fever—a systematic review of primary and secondary cases. J Travel Med

Yasri S, Wiwanitkit V (2018) Magnitude of renal problem in fatal cases of Lassa fever: a summary on existed reports. Saudi J Kidney Dis Transpl Official Publ Saudi Center Organ Transpl Saudi Arabia 29:1518

Yun NE, Ronca S, Tamura A, Koma T, Seregin AV, Dineley KT, Miller M, Cook R, Shimizu N, Walker AG, Smith JN, Fair JN, Wauquier N, Bockarie B, Khan SH, Makishima T, Paessler S (2015) Animal model of sensorineural hearing loss associated with Lassa virus infection. J Virol 90:2920–2927

Yun NE, Walker DH (2012) Pathogenesis of Lassa fever. Viruses 4:2031–2048

Zapata JC, Cox D, Salvato MS (2014) The role of platelets in the pathogenesis of viral hemorrhagic fevers. PLoS Negl Trop Dis 8:e2858

MIX
Papier aus verantwortungsvollen Quellen
Paper from responsible sources
FSC® C105338

If you have any concerns about our products,
you can contact us on
ProductSafety@springernature.com

In case Publisher is established outside the EU,
the EU authorized representative is:
Springer Nature Customer Service Center GmbH
Europaplatz 3, 69115 Heidelberg, Germany

Printed by Libri Plureos GmbH
in Hamburg, Germany